TAYLOR & FRANCIS MONOGRAPHS ON PHYSICS

EDITOR
B. R. COLES, D.Phil., B.Sc.
Professor of Solid State Physics, Imperial College of Science and Technology, London

CONSULTANT EDITOR
SIR NEVILL MOTT, F.R.S.
formerly Cavendish Professor of Experimental Physics, University of Cambridge

DEFECTS IN THE ALKALINE EARTH OXIDES

DEFECTS IN
THE ALKALINE EARTH OXIDES
with applications to radiation damage and catalysis

B. HENDERSON

Trinity College, Dublin, Ireland

and

J. E. WERTZ

University of Minnesota, Minneapolis, U.S.A.

TAYLOR & FRANCIS LTD
LONDON

HALSTED PRESS
division
John Wiley & Sons Inc.
NEW YORK—TORONTO

1977

First published in ADVANCES IN PHYSICS, *Volume 17, No. 70, November 1968.*
This revised edition published 1977 by Taylor & Francis Ltd, London and Halsted Press (a division of John Wiley & Sons Inc.), New York.

© 1977 Taylor & Francis Ltd.

All rights reserved. No part of this publication may be reproduced, stored in a retrieval system, or transmitted, in any form or by any means, electronic, mechanical, photocopying, recording, or otherwise, without the permission of Taylor & Francis Ltd.

Taylor & Francis ISBN 0 85066 086 6

Printed and bound in Great Britain by Taylor & Francis (Printers) Ltd, Rankine Road, Basingstoke, Hampshire.

Library of Congress Cataloging in Publication Data

Henderson, B.
    Defects in the alkaline earth oxides.
    (Taylor & Francis Monographs on Physics)
    First published in Advances in Physics, Volume 17, No. 70.
    Bibliography: p.
    Includes index.
    1. Alkaline earth oxides—Defects. I. Wertz, John E., 1916–   joint author.
    II. Title.
QD921.H46   1977   546'.39   77–23366
ISBN 0–470–99205–0

Preface

The study of colour centres in alkali halide crystals began in Europe some five decades ago. Since that time, as a result of very many investigations, the symmetry and structure of numerous defects have been unravelled. The reason for these studies was that alkali halide crystals are model systems in which to probe, both experimentally and theoretically, many aspects of the defect solid state. In addition, large single crystals of high purity may be made at rather modest cost. Study of the isostructural alkaline earth oxides formed a natural extension of such work. Nevertheless work on the oxides is a comparatively short-lived phenomenon. Problems associated with impurity levels and the commercial availability of large transparent single crystals have made definitive study frustrating. However, in 1968, there did appear to be a considerable body of information which merited an exhaustive discussion of the defect structure of the alkaline earth oxides. In consequence the authors published a lengthy paper, 'Defects in the alkaline earth oxides', in the review journal *Advances in Physics*. The present monograph is an outgrowth from that review paper. It would have been a far simpler task to correct the few errors in the original version and re-issue it between hard covers. Fortunately this area of research activity has continued in the intervening eight years to make rapid progress, such that the original would in no way have been adequate. In consequence a major overhaul seemed in order: the current volume is the result of such an overhaul. Some 60 percent of the original review remains in more or less unexpurgated form. However, the subject matter has been substantially rearranged in accommodating the newer material. In general this monograph is written for scientists actively working in the physics and chemistry of the defect solid state, although to make the account more tutorial and therefore of more value to senior undergraduates and beginning postgraduates, a chapter has been added on the simple theory and practice of some techniques used in the study of colour centres.

The current volume was undertaken by one of the authors of the original review (B.H.) and he alone is responsible for any errors of omission or fact. It could not have been achieved without the encouragement and stimulus of very many friends. In this context, we owe a special debt of gratitude to Professors D. L. Cowan, A. H. Garrison, D. J. E. Ingram, J. C. Kemp, Y. Merle d'Aubigné and W. A. Sibley as well as to Drs M. M. Abraham, Y. Chen, J. J. Davis, A. E. Hughes, A. M. Stoneham and R. F. Wood. We are also grateful to these and others such as Drs R. D King, W. Low, I. K. Ludlow, J. H. Lunsford and D. J. Miles, for permission to use original figures.

B. Henderson
Dublin
October 1976

Contents

PREFACE   v

1. INTRODUCTION TO DEFECTS AND THE ALKALINE EARTH OXIDES
   1.1. Preamble   1
   1.2 Crystal growth and macroscopic impurity effects   2
   1.3. The effects of annealing on the perfection of single crystals   6
   1.4. Defects in ionic crystals   7
   1.5. Defect creation in the alkaline earth oxides   8
      1.5.1. Additively coloured crystals   8
      1.5.2. Radiation colouration in the alkaline earth oxides   11

2. PRINCIPLES OF MAGNETIC RESONANCE AND OPTICAL SPECTROSCOPY   13
   2.1. Basic features of electron spin resonance   13
      2.1.1. The spin Hamiltonian   16
   2.2. Optical spectroscopy of solids   20
      2.2.1. The configurational coordinate model   22
      2.2.2. Uniaxial stress effects   25
   2.3. Experimental considerations   27

3. OPTICAL AND MAGNETIC PROPERTIES OF IMPURITY CENTRES   30
   3.1. Electron spin resonance spectra of transition metal ions   31
      3.1.1. Forbidden lines in impurity spectra   45
      3.1.2. Unusual aspects of impurity spectra in MgO   45
   3.2. Effects of heat treatment   50
   3.3. Effects of ionizing radiation   50
   3.4. Luminescence properties of impurity ions   51

4. TRAPPED HOLE CENTRES IN OXIDE LATTICES   56
   4.1. Effects of ionizing radiation. V centres   56
   4.2. Formation and stability of V centres in MgO   59
   4.3. The nature of trapped hole centres   61
   4.4. Modified V centres   63
      4.4.1. The V_{OH} and V_{OD} centres   63
      4.4.2. The V_D centre   66
      4.4.3. Other centres   67
   4.5. Optical absorption of trapped hole centres   68
   4.6. More on the nature of trapped hole centres   71

5. THE STRUCTURE AND PROPERTIES OF ELECTRON EXCESS CENTRES   75
   5.1. The ESR and ENDOR spectra of F^+ centres   76
   5.2. Optical properties of F^+ centres   82
      5.2.1. Magnesium oxide   83
      5.2.2. Calcium oxide   86
      5.2.3. Strontium and barium oxides   86

5.3. Magneto-optic and vibronic properties	87
5.4. Luminescence from F^+ centres	93
5.5. F centres	95
5.6. Theoretical studies of F and F^+ centre properties	98

6. THE STRUCTURE AND PROPERTIES OF DEFECT AGGREGATES 102
   6.1. F_A^+ centres in calcium oxide 103
   6.2. Exchange-coupled defect clusters 104
   6.3. Magnetic resonance studies of annealed crystals 108
      6.3.1. Vibrational structure 113
   6.4. Symmetry of colour centres in magnesium oxide 115
      6.4.1. Orthorhombic centres 116
      6.4.2. Trigonal centres 119
      6.4.3. Emission studies of trigonal centres 125
   6.5. Other studies in magnesium oxide and calcium oxide 125

7. THE NATURE OF IRRADIATION DAMAGE IN MAGNESIUM OXIDE 126
   7.1. Evidence of point interstitials 128
   7.2. Annealing studies and cavity formation 131

8. OXIDE SURFACES AND CATALYSIS 133
   8.1. ESR studies of defects in and on polycrystalline oxides 133
      8.1.1. Formation of F^+ centres in oxide powders 135
      8.1.2. Trapped hole centres in magnesium oxide powders 136
      8.1.3. Trapped hole centres at surfaces 136
      8.1.4. Adsorbed paramagnetic species on magnesium oxide powders 137
   8.2. Optical studies of oxide surfaces 140
   8.3. Heterogeneous catalysis and chemical reactions on magnesium oxide surfaces 142

REFERENCES 144

AUTHOR INDEX 153

SUBJECT INDEX 157

DEFECTS IN
THE ALKALINE EARTH OXIDES

1. INTRODUCTION TO DEFECTS AND THE ALKALINE EARTH OXIDES

1.1. *Preamble*

Investigations of the defect solid state have been actively pursued on a world-wide basis for several decades. Initially the end products of such studies had little or no application, but as in many other areas of fundamental research technological benefits have accrued in the long term. Using the term *defect* to encompass impurities in crystals the following list of applications, infra-red quantum counters, radiation dosimeters, photochromic computer memory devices, electron beam memory devices, display tubes, is not exhaustive. However, the increasing use of these and similar applications of the properties of defects offers a considerable justification of the capital expenditure invested in such fundamental research. The present treatise is concerned not with technology but with the underlying physical principles applied in some investigations of the defect solid state.

Studies of defects in non-metallic solids have concentrated on the alkali halides and the elemental semiconductors silicon and germanium. The reasons for studying the alkali halides are partly historical. However, the availability of large, pure single crystals has been especially influential. In addition the simple crystal structure and essentially ionic bonding make the alkali halides ideal insulating host lattices in which to study a wide variety of defect species. The impetus for the research on silicon and germanium has been technological, and single crystals may be obtained with ease and in a greater state of purity and perfection than even the alkali halides. Thus, in both types of solids, those physical and chemical properties which are strongly influenced by crystalline imperfections including excitons, vacant and interstitial lattice sites, impurities and dislocations, have been investigated to an almost unparalleled extent. It is no surprise that for these solids, measurements of transport phenomena, optical properties and magnetic resonance have been responsible for the present understanding of the defect state (Seitz 1954, 1946, Schulman and Compton 1962, Fowler 1968, Corbett 1966).

The alkaline earth oxides, beryllium oxide excepted, are the divalent structural analogues of the alkali halides. Thus these crystals are face-centred cubic, the anions and cations being situated at the corners of two interpenetrating cubic sublattices. Despite being highly ionic and having the simple rocksalt crystal structure, the alkaline earth oxides have not been as extensively investigated as have the alkali halides. Their high melting points result in the presence of large dislocation densities and impurity concentrations in single crystals which have usually been prepared by the arc fusion method. Such imperfections greatly hinder the understanding of phenomena associated with the presence of intrinsic lattice defects in the oxides. Only recently have the optical properties of the transition group impurities become well understood. Consequently a much better understanding of the properties of intrinsic lattice defects, transport phenomena and radiation effects in these oxides has now been achieved. Thus it is apposite to review the present status of our knowledge of defects in these insulating solids.

We survey the progress of the last two decades. Conveniently we start by discussing the purity of presently available single crystals; the impurity distribution is intimately connected with the mosaic structure of even the best single crystals. Some impurities are present mainly as constituents of precipitates formed during the cooling process. Others are dissolved substitutionally in the lattice, and optical absorption and fluorescence as well as electron spin resonance studies have given considerable insight into the behaviour of these impurities. Other defects have also been extensively investigated, and the recent progress in this field is discussed in relation to the radiation damage processes. It will be seen to parallel closely the numerous recent developments in the alkali halides. The formation of macroscopic defects, dislocation loops and macroscopic voids in neutron irradiated samples is also discussed. Finally, we relate the properties of surface defects, which have been the subject of a very careful study recently, to the processes of chemisorption and physisorption of gaseous species on the surfaces of powder samples. The term 'defect' has been interpreted very loosely by the authors: even so, the properties of dislocations are discussed only briefly in so far as they are necessary to the study of impurity and radiation effects. The mechanical properties of magnesium oxide have been extensively reviewed by Miles (1964).

1.2. *Crystal growth and macroscopic impurity effects*

The high melting temperatures of the alkaline earth oxides preclude the use of elegant techniques of single crystal growth and purification that are commonplace to scientists working with semiconductors and the alkali halides. The universal technique used for the oxides is that of arc fusion. A large mass of oxide, hydroxide or carbonate powder is packed around carbon electrodes and an arc is struck between them. The high temperature in the arc is sufficient to melt the powder, the large mass of which ensures that the molten pool cools slowly. Crystals of varying size and purity are obtained. The sizes of crystals are increased when the powder is compacted to about 80% theoretical density and if three or four-electrode systems are utilized. Magnesium oxide single crystals, several cubic inches in volume, may be obtained from various organizations. Smaller crystals of calcium oxide, strontium oxide and barium oxide produced by this technique are also available.

Three other techniques have been applied to the growth of single crystals of the alkaline earth oxides. Small platelets of magnesium oxide have been grown by pulling from a lead molybdate flux. The crystals contained relatively large inclusions from the melt and also had large concentrations of grown-in dislocations. Gambino (1965) produced large single crystal boules of strontium oxide and barium oxide using a plasma torch method. This method seems potentially useful for doped single crystals since Gambino incorporated detectable concentrations of Eu^{2+} into both oxides. The paramagnetic resonance spectra of Eu^{2+} in both materials indicates that the site symmetry around the ions is essentially octahedral. It is inevitable that the very rapid cooling rates involved in this process will cause the development of substantial mosaic structure in the crystals. The use of after-heaters to control the cooling rate seems essential before this technique can become more generally useful. MgO crystals 2 mm on edge may be grown by the flux evaporation technique (Garton et al., 1972), a method useful for both ^{17}O enrichment and impurity doping.

Unfortunately, no purification techniques are currently available for these refractory solids; hence it is essential to use starting powder of the highest possible purity. This problem was not readily recognized by the early manufacturers, with the result that many investigations were concerned with impurity effects. Some typical analysis of commercially available single crystals are given in table 1. It is notable that magnesium, calcium and strontium oxide crystals all contain varying concentrations of hydroxyl ions (detected by infra-red absorption) in addition to the metal ion impurities. The large concentrations of transition metal ions are of importance since irradiation may change their valence states as well as producing intrinsic lattice defects. The impurity valence changes may completely mask the purely intrinsic effects, and for this reason it may be difficult to draw confident conclusions about the nature and effects of the radiation damage, unless high resolution techniques such as ESR are used.

The presence of non-transition metal ions may have less effect upon the events occurring during irradiation. In magnesium oxide such impurities form compounds which may precipitate out during cooling to decorate the dislocation lines which are observed in the optical or electron microscope (Venables 1963, Bowen 1963, Bowen and Clarke 1963). Typical micrographs of dislocation networks in magnesium oxide and calcium oxide are shown in fig. 1. The precipitation of impurities at dislocation lines makes observable a network of low angle boundaries; the latter separate sub-grains ranging in diameter from 0·5 to 2·0 mm and which may involve misorientations of a few minutes of arc. Particle diameters up to 0·5 μ are common, and separation between precipitates along the dislocation lines of order 0·1–10 μ are clearly evident in electron micrographs.

The chemical composition, size and distribution of these precipitates has important consequences in respect of the microstructure and strength of magnesium oxide (Stokes 1962, Miles et al. 1965, Davidge 1967). Venables (1963) showed that the majority of precipitates in 'as-received' magnesium oxide are zirconium dioxide, despite there being only about 1 p.p.m. of zirconium in the crystals and considerably larger concentrations of silicon, calcium, aluminium and iron which also might be expected to form precipitates. The geometrical forms of these precipitates have been discussed by Matkin and Bowen (1965): both blade-like and needle-like precipitates were identified with zirconia. Circular platelets were shown to be composed of cubic lime-stabilized zirconia while faceted platelets were composed of monoclinic zirconia. The blade-like and needle-like precipitates have different habit planes and growth directions, the actual shape being determined by the structural and bond interfacial energies and the strain energy in the matrix resulting from a difference in volume between the two phases. In view of the differences in shape of the cubic stabilized zirconia and the monoclinic structure it would appear that the interfacial energies

Table 1. Analyses of some oxides, including powders and commercially available single crystals

Impurity	'Specpure' MgO powder†	'Kanto' MgO powder	Norton single crystal MgO	'Spicer' single crystal MgO‡	'Spicer' single crystal CaO	'Spicer' single crystal SrO
Al	30–50	50	100–200	35	70	10
Ca	20	15	300–500	20		70
Si	15–20	40	10–50	15	30	20
Fe	5–20	5	30–200	3	20	15
Ni		5	2		2	
Mn	1	3	5	0·1	10	5
P		7		2		
Pb	2–3	<1	<2	<1		7
S		50		5		
Sn	<1					2
Cu	<5	<2	<5	<1	<1	
Sr		<1	<1		5	
Ge	1–5				2	
Mo			<2			
Zn	<20	20	<50	5	40	30
Cd, Tl, Pd	<2		<1		<7	
V	<10	<10	30	<2	10	5
Mo	<5		<2			
Cr	<10	1	15	<1	<17	2
As, K	<20	<5		<5		
Bi			<1			
Ti	<50	<20	<20	<20		
Rb	<50					2
Be			<1			
Co	7	<1	<2			
Sb		<1	<5		15	7
Ba		<1	<10	<0·5	2	15
Na	20	20		1		
Zr		<1	<20	3	20	

  † Obtained from Johnson-Matthey Co. Ltd.
  ‡ Produced by W. & C. Spicer Ltd. from powder obtained at the Kanto Chemical Co. Ltd.

Fig. 1

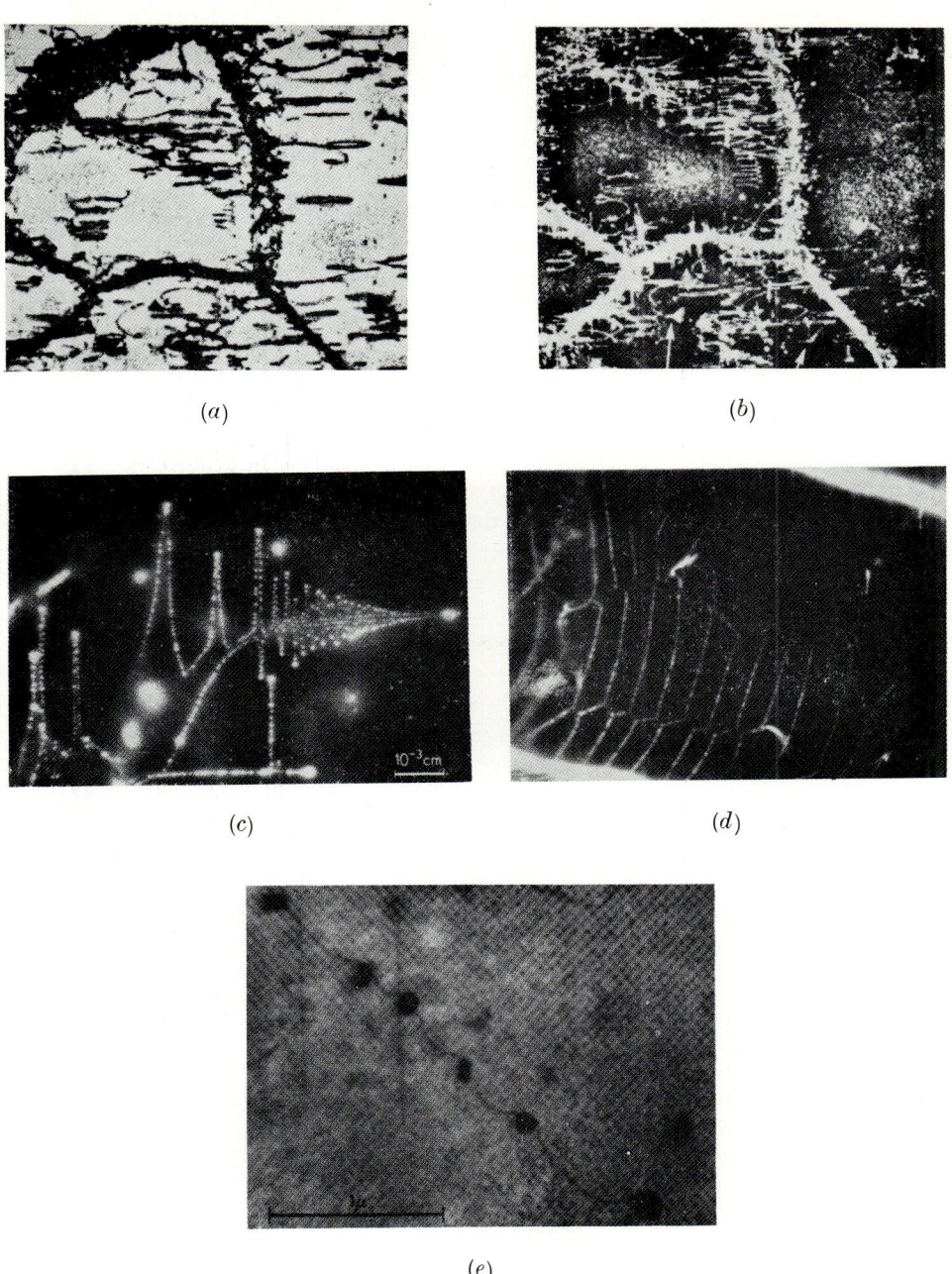

Microstructures of MgO and CaO, (*a*) and (*b*) are respectively mosaics of X-ray topographs and ultramicrographs of same area (after Miles and Lang 1965), (*c*) ultramicrograph of an unusual dislocation configuration in MgO (after Miles 1964), (*d*) sub-grain boundaries in CaO seen in dark field and (*e*) Transmission electron micrograph of precipitates on dislocation lines in MgO (after Miles 1964).

are most important in determining the shape of precipitates. Other precipitates including spinel, magnesium hydroxide and magnesio-ferrite have been identified using electron diffraction or electron probe micro-analysis techniques (Henderson 1964, Briggs and Bowen 1968, Groves and Fine 1964).

1.3. *The effects of annealing on the perfection of single crystals*

Frequently the single crystals of magnesium oxide, calcium oxide and strontium oxide grown by the arc fusion process have a cloudy appearance; this appears to be due to the presence of voids and macroscopic impurity distributions in the crystals. Good optical quality is obtained by annealing the crystals at temperatures of 1800–2000°C in high vacuum (10^{-5}–10^{-6} torr) for several hours. Optical microscope investigations have shown that such heat treatments also affect the impurity precipitate distribution quite markedly, at least in magnesium oxide (Bowen 1963, Bowen and Clarke 1963). These studies demonstrated that precipitates are revealed by the presence of pyramidal features on cleaved magnesium oxide surfaces which had been etched in hot orthophosphoric acid or fuming nitric acid. The distribution of pyramid sizes is believed to represent the distribution of precipitate sizes (Bowen and Clarke 1963). Thus by studying these pyramidal features in annealed and quenched samples it has been shown that the precipitate concentration remains almost unaffected in crystals quenched from about 800°C: in the range 800–2000°C a progressive decrease is observed in the number and size of precipitates not associated with the presence of grown-in dislocations. The observations are consistent with a dissolution of many of the impurities as a result of the heat treatment. Those precipitates associated with dislocations appear to dissolve much less readily than those individually distributed throughout the crystal. At temperatures above 1800°C almost complete clearance of the precipitates was observed in a surface layer the thickness of which depended upon the annealing time. Bowen and Clarke (1963) suggested that a diffusion mechanism with a diffusion coefficient at 1900°C of about 10^{-7} cm^2 sec^{-1} was responsible for the denuded region. The diffusing species was not identified.

Magnesium oxide crystals annealed at very high temperatures attain extremely high maximum shear strengths. After heat treatment for $\frac{1}{2}$ hour at 2000°C in an argon atmosphere Stokes (1962) observed maximum shear strengths greater than 5×10^9 dyn cm^{-2}. To explain these high strengths, Stokes proposed that precipitates in as-received single crystals acted as dislocation sources, and that heat treatment removes the precipitates. In fact, even the highest temperatures did not remove all the precipitates, except perhaps in parts of the denuded surface layers. Thus, the high strengths cannot result from the complete removal of precipitates. Alternative acceptable explanations might involve dislocations lying on slip planes unfavourable to movement, or the pinning of dislocations by small impurity particles or by an impurity atmosphere.

Heat treatment is of great importance to the valence state of transition metal ions which substitute for Mg^{2+} ions on the cation sublattice. When carried out in high vacuum or in argon, these treatments reduce most transition metal ions present in the alkaline earth oxides to the divalent state. Strongly reducing atmospheres such as hydrogen or carbon monoxide produce similar effects. As discussed in § 7 such reducing atmospheres also introduce large, hydrogen-containing cavities of rectangular cross-section into the crystals. The hydrogen gas is at high pressure and is a consequence of the hydroxide content of the crystal rather than the diffusion of hydrogen into the crystal. If the annealing treatments are carried out in air, then some of the transition metal ions change to a higher oxidation state, which may readily be detected by paramagnetic resonance or optical absorption studies. Such effects are dealt with in greater detail in Chapter 3.

1.4. *Defects in ionic crystals*

There are many categories of defects in ionic crystals, the nomenclature of which has developed in rather haphazard fashion. Consequently the recent attempt by Sonder and Sibley (1972) to introduce a consistent nomenclature applicable to all polar crystals is particularly laudable. Their system is adhered to here. The simplest defects are impurity ions, and such defects are abundant in all commercially available oxides. A substitutional impurity is denoted as $[Y^-]$ or $[Me^{2+}]$, where Y and Me denote anion and metal ion respectively and the valence state is represented by the superscript. Thus in KCl where thallium impurity has been much studied the electronic defects $[Tl^0]$ and $[Tl^{2+}]$ may be produced during ionization radiation. In the alkaline earth oxides the transition metal ions change their valence state rather easily and so produce a variety of electronic defects. For example, titanium has been observed as Ti^+ and Ti^{2+} in CaO and iron as Fe^+, Fe^{2+} and Fe^{3+} in both MgO and CaO. The variety of electronic defects on the anion sublattice is more limited, although the hydroxyl and fluoride ions may be present both substitutionally and in association with other defects.

Under certain circumstances vacant lattice sites and interstitial ions are observed in the alkaline earth oxides: these are *lattice* or *ionic* defects. Vacancies may occur on either anion or cation sub-lattice. Anion vacancy centres are distinguished by the letter F. In divalent compounds there are several possible defect species. The anion vacancy may trap zero, one, two or perhaps three electrons. To differentiate between these species the charge of the defect relative to the perfect lattice is indicated by a superscript. Thus the anion vacancy containing a single electron is referred to as an F^+ centre, whilst the addition of a second electron results in the F centre. So that there is no confusion with centres in alkali halide crystals we emphasize that the term F centre is used to indicate *an anion vacancy which is uncharged with respect to the lattice*. When the anion vacancy centre is located adjacent to an impurity centre the aggregate defect is referred to as an F_A centre. For example in oxide crystals, F^+ centres adjacent to Me^{2+} ions are referred to as F_A^+ centres. Since the nature of the impurity may change most authors also indicate the interacting impurity species. Thus the above centres are alternatively referred to as the F_{Me}^+ centre or as the $F_A^+(Me)$ centre.

Vacancies which occur on the cation site are indicated by the letter V. Accordingly the V^- centre indicates a cation vacancy, adjacent to which is a single positive hole shared by the near-neighbour oxygen ions. When two positive holes are localized on the oxygen ions neighbouring the cation vacancy the V centre results. Modified V-type centres are observed in impure crystals; these have been labelled V_F, V_{OH} or V_{Al} according as the nearby impurity is F^-, OH^- or Al^{3+}. These three centres are all neutral in charge with respect to the lattice. Some simple single-vacancy centres are illustrated in fig. 2. The Sonder–Sibley system is easily adapted to aggregates of vacancies by use of a subscript to indicate the number of vacancies in a particular aggregate. Thus two nearest neighbour F centres would be referred to as the F_2-centre. There are many variants on the structure of the aggregate centres in the various ionic crystals. For particular details the reader should consult the paper by

Sonder and Sibley (1972). Further development of the nomenclature will be explained as appropriate in the text.

Fig. 2

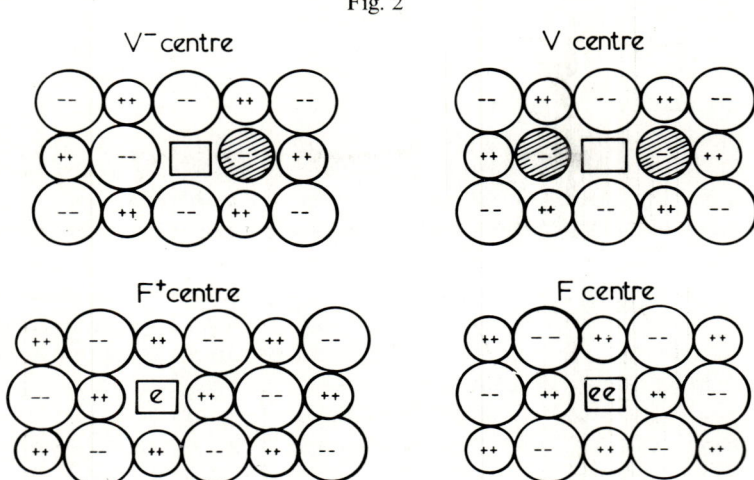

Structure of single anion and cation vacancy centres in the alkaline earth oxides.

Many ionic crystals are transparent to photons with wave lengths in the ultraviolet, visible and near infra-red regions. Most defects introduce energy levels into the band gap of such crystals and so cause selective absorption of some components of the visible spectrum: the crystals are then coloured. The term *colour centre* was introduced by the very early investigators to describe the defects responsible for the coloration. Originally *colour centres* described only the intrinsic ionic defects in alkali halide crystals: today the term embraces all defects which colour insulators, including impurities.

1.5. *Defect creation in the alkaline earth oxides*

Colour centres are generated in the alkali halides by a variety of techniques, including additive coloration, electrolytic coloration and radiation coloration. These techniques, especially excess cation coloration and radiolytic coloration have also been used in the alkaline earth oxides. Unlike the alkali halides, trapped electron centres are not produced by X-irradiation, except when the anion vacancies are generated by plastic deformation priot to X-irradiation (Wertz *et al.* 1957; 1962, Bessent and Feltham (1968). Such defects have also been observed in explosively shocked magnesium oxide without X-irradiation (Gager *et al.* 1964). However, controlled concentrations of defects cannot be generated in this way and the resulting crystals are considerably strained.

1.5.1. *Additivily coloured crystals*

An excess of either the alkaline earth metal or the oxygen constituent may in principle be introduced into the crystals by heating to a high temperature in the vapour of the constituent. Excess metal coloration is successful in magnesium oxide (Weber 1951), calcium oxide (Ward 1965, Ward and Hensley 1965, Kemp *et al.* 1967 b) and barium oxide (Dash 1953, Carson *et al.* 1959). Attempts to introduce excess oxygen into magnesium oxide (Soshea et al. 1958, Wertz et al. 1959) and calcium oxide (Tomlinson and Henderson 1968) result principally in changes in the valence state of impurities in the crystals. The only systematic attempt to additively colour the alkaline earth oxides has been by Hensley and his colleagues at the University of Missouri (Ward and Hensley, 1968, Hensley et al. 1968, Johnson and Hensley 1969).

Excess alkaline earth metal ions in the oxides require simultaneous incorporation of a corresponding number of electrons elsewhere in the crystal, if the system as a whole is to remain electrically neutral. The electrons may be trapped in the field of the excess ions, M^{++}, if these ions occupy interstitial sites. It is more probable that the excess cations are substituted on normal lattice sites, simultaneously with the formation of oxygen ion vacancies in which the electrons are trapped. The anion vacancy may trap one, two or perhaps, three electrons, and the optical absorption spectrum from crystals containing all three such defects may not readily be interpreted. For these three oxides the absorption spectra produced by coloration with excess of a different cation (e.g. Ca in BaO) is substantially the same as that obtained from a stoichiometric excess of the constituent cation.

Weber (1951) first additively coloured magnesium oxide with excess cation: coloration was carried out in steel tubes at 1150°C but no precautions were taken to ensure rapid cooling of the samples after additive coloration. Wertz et al. (1962) have commented that the absorption bands present in additvely coloured crystals are accompanied by Tyndall scattering. Although this scattering has not been examined in detail, comparisons with studies of the alkali halides (Seitz 1954) suggest that at least part of the absorption is due to colloidal centres. Such centres arise from an association of F centres to form large agglomerates. Certainly the bandwidths reported by Weber (1951) suggest this to be the case since he observed a width of 1·1 eV for the 4·90 eV band. This is very much broader than the band subsequently observed by Hensley and Kroes (1968) and the F^+ band in neutron irradiated crystals (see § 5.2). It appears that Weber's quenching rates were too slow to preclude the agglomeration of F centres into colloids. If the crystals are rapidly cooled, three bands at similar wavelengths to those obtained by Weber are obtained but with much reduced halfwidths. The actual energies at which the bands are observed at 300°K are 4·90 eV, 4·07 eV and 2·20 eV. A band at 4·90 eV in neutron irradiated crystals has been identified with the F^+ centre (Wertz et al. 1964, Kemp and Neeley 1963 a, Henderson and King 1966, Kemp et al. 1966). Work on additively coloured, neutron irradiated and 1·6 MeV electron irradiated crystals suggest that the F band also occurs at this wavelength. Thus the absorption at lower energies must be associated with defects other than F^+ and F centres. It is not unlikely that anion vacancies may trap three electrons to form F^- centres and such centres could maintain electrical neutrality compensated by the presence of F^+ centres. One or other of the bands at 4·00 and 2·2 eV might then be due to such a centre. This latter suggestion is certainly speculative, and it may be more reasonable to suppose that the bands are due to aggregate defects or interstitials.

Several studies of additively coloured calcium oxide have been reported recently (Ward 1965, Ward and Hensley 1965): single crystal samples were heated at 1750° in one to several atmospheres of calcium vapour and rapidly cooled. The crystals were then annealed, presumably to dissociate colloidal centres. A spectrum similar to that shown in fig. 26 (b) results, in which the F^+ band (3·65 eV band) and the band at 3·1 eV are clearly resolved. (The F^+ band has been identified using magnetic resonance, optical absorption and Faraday rotation studies by Kemp et al. (1967 a).)

The band at 3·1 eV is believed to be the F band (Ward 1965, Ward and Hensley 1965), and there is now available evidence consistent with this view. Kemp, Ziniker and Hensley (1967) have reported electron transfer between F^+ and F centres under photoexcitation conditions, as measured by changes in the absorption intensities of the 3·1 eV and 3·65 eV bands and of the F^+ centre spin resonance signal. With photons of energy greater than 2·8 eV, the F^+ band (3·65 eV) intensity increased and the F-band (3·1 eV) intensity decreased, with approximate sharing of the total intensity between the bands. For light with energy less than 2·5 eV the F band increased at the expense of the F^+ band. These observations demonstrate that during high energy photon irradiation the F centre loses one electron, thus increasing the F^+ centre concentration. The excited electron becomes trapped at some other site in the lattice from which it may be re-excited to repopulate the F centres during irradiation with 'red' light. These reversible changes observed by Kemp et al. (1967 b) at 77°K are essentially similar to results obtained at 300°K but with different transient behaviour. At 4°K no changes were observed using optical or spin resonance methods with either 'blue' or 'red' light. To explain these observations, Kemp et al. (1967 b) proposed that the 'blue' light ($hv > 2·8$ eV) raised the electron to an excited state ($^1S \rightarrow {}^1P$). It may then be thermally excited into the conduction band from whence it is trapped at other levels in the lattice. Thus the F^+ band and the F^+ centre spin resonance signal increase at the expense of the F centre when the experimental temperature is high enough to assist the electron excitation process. At 4°K there is insufficient thermal energy for these excitation processes typical of the behaviour at 300°K and 77°K to take place. During irradiation at the higher temperatures with 'red' light the F^+ band intensity decreases due to the F^+ centres trapping electrons released during irradiation. The nature of the defect species from which these photo-electrons are released is obscure, but the numerous impurities present in the crystals (e.g. Fe^{2+}, Mn^{2+}, V^{2+}, etc.) are obvious possibilities.

Single crystals of strontium oxide may be additively coloured by heating them in strontium vapour at pressures of 0·35–42 torr in the temperature range 166–196°K. Both F^+ and F centres are produced by this treatment (Johnson and Hensley 1969).

Sproull et al. (1951) first reported the growth of high purity single crystals of barium oxide by sublimation onto magnesium oxide seed crystals. The absorption spectra were obtained after coloration with excess calcium, magnesium or aluminium and are substantially the same as that obtained from a stoichiometric excess of barium, all the curves show a prominent peak at 2·0 eV (Sproull et al. 1953). This peak is due either to interstitial barium or to oxygen ion vacancies containing trapped electrons. Sproull et al. (1953) measured the diffusion of the coloured region into the single crystal sample over the range 800–1300°C and observed a unique activation energy of 2·8 eV. Activation energies of 11·0 eV and 0·4 eV have been reported for the diffusion of radioactive ^{140}Ba into barium oxide (Redington 1952). It would seem, therefore, that in additively coloured barium oxide, the absorbing species are the F-like centres. Carson et al. (1959) observed the ESR spectrum from the F^+ centre in some additively coloured samples, but could not correlate this spectrum with the 2·0 eV band. That the band at 2·0 eV is due to the F^+ centre has been demonstrated in proton irradiated barium oxide by Bessent et al. (1968) using the paramagnetic Faraday rotation technique. These authors also suggest that the high energy tail in the F^+ band results from the first excited state of the F^+ band in barium oxide being in or very close to the conduction band.

1.5.2. *Radiation coloration in the alkaline earth oxides*

The highly ionic alkali halides are extremely susceptible to damage by ionizing radiation. It is well established that the production of radiation defects in these crystals is extremely sensitive to electron–hole recombination processes. Since these processes are produced primarily by ionizations, they are low-energy events. In the alkaline earth oxides, the higher valencies of the ions lead to much larger binding energies than in the alkali halides. There is also a larger covalent contribution to the binding. Thus one anticipates the radiation damage process to involve much larger displacement energies.

In the alkali halides where low-energy displacement processes are involved, a particularly simple mechanism of damage by ionizing radiation is envisaged (Pooley 1966). The mechanism involves the non-radiative decay of the excited excitonic states of the crystal. All the energy absorbed in exciton production is transferred to an atom and may be dissipated in suitable cases by a replacement collision sequence along a line of anions. This [110] directed collision cascade may be a central feature of the F-centre production process. Pooley used a computer simulation of the radiation damage to calculate the energy required to create a stable vacancy–interstitial pair in the [110] line of anions. Pooley's (1966) earliest calculations gave [110] replacement thresholds of about 4–5 eV except where the anions are too closely packed. More realistic two and three-dimensional models yield [110] much higher replacement thresholds (Hughes *et al.* 1967, Torrens *et al.* 1966) and apparently further investigations of the choice of the lattice potential are in order.

Similar calculations give threshold energies in excess of 40 eV for a [110] replacement collision cascade in magnesium oxide (Pooley 1966), and indicate with reasonable certainty that ionization processes do not cause ionic displacement in the alkaline earth oxides. In conformity with these ideas, Wertz *et al.* (1957, 1961, 1962) have reported that F^+ centres are not observed in magnesium oxide after irradiation with ultra-violet light, X-rays and γ-rays. Sibley and Chen (1967) and Chen *et al.* 1970 have investigated the colorability of magnesium oxide during electron irradiation as a function of electron energy, radiation intensity and radiation dose. They observe that the intensity of the F^+ band† for a given dose increases with increasing radiation energy for electrons with energy greater than 0·7 MeV. However, electron irradiation at energies below 0·33 MeV produces few or no F^+ centres. Since relativistic particles may transfer to an atom a maximum kinetic energy given by:

$$E_{\max} = \frac{2E(E + 2mc^2)}{Mc^2}, \quad (1)$$

where E and m are the energy and mass of the electron, c is the velocity of light and M is the energy of the struck atom, the displacement energy for oxygen ions in magnesium oxide is apparently about 50 eV. This is perhaps fortuitously close to the rough value obtained by Pooley (1966) for the threshold energy of a [110] replacement cascade event. Sibley and Chen (1967) also reported that the F^+ centre production rate is somewhat insensitive to electron irradiation intensity, and that for intensity variations over a factor of 3 the F^+ centre concentration depends only on

---

† Chen *et al.* (1968 a) have shown that in electron irradiated MgO both F and F^+ centres are present, although F centres are predominant. This does not alter the present arguments related to the displacement energy, since both are anion vacancy centres.

the radiation dose. If damage by ionization is involved the damage rate is expected to be insensitive to variations in radiation energy and intensity. Thus the results for magnesium oxide indicate energy transfer by elastic collisions.

Less work has been done on the other alkaline earth oxides. However, work on both single crystals and powder specimens of calcium oxide and strontium oxide shows that F^+ centres are not generated during irradiation with ultra-violet light, X-rays or 400 kV electrons. High-energy proton irradiation (2–10 MeV) does produce F^+ centres, as demonstrated by the recent magnetic resonance, optical absorption and Faraday rotation studies of Bessent et al. (1968). These results support the conclusion reported above for magnesium oxide and suggest that neutron irradiation should be especially useful for producing displacement damage in the oxides. In fact reactor irradiation has been used most frequently to study colour phenomena and radiation damage in the oxides.

A nuclear collision results in the neutron being either scattered away or absorbed by the nucleus. The former process results in a sharing of energy between the nucleus and neutron, and the probability of such an event occurring is measured by the differential scattering cross section σ_s of the atom. The latter process involves transmutation and is of some importance in beryllium oxide and magnesium oxide since it produces inert gas atoms which may exist in large cavities in the irradiated crystals (§ 7).

The neutron has no charge and hence it does not exert an appreciable force on an atom, except when it strikes the nucleus. Most such collisions are elastic, and the energy imparted to the atom in the process ranges from 0 to E_{max}, where the maximum energy E_{max} transferred in a head-on collision is related to the incident neutron energy (E) by:

$$E_{max} = \frac{4mM}{(m+M)^2} E, \qquad (2)$$

where m and M refer to the masses of the neutron and atom respectively. This energy may be dissipated in a number of ways as discussed below. Since the displacement energy for the oxygen ion in the alkaline earth oxides is of order 50–100 eV, it is apparent that all fast neutrons can produce displacements. In a material of atomic density N, the number n_p of atoms undergoing scattering collisions in a given integrated flux ϕ of neutrons is given by $n_p = N\sigma_s\phi$. If the energy of the primary event is greater than a certain value L_e, then the energy is lost by the ionization of atoms along the path of the primary displacement and does not produce displacements. For energies below the ionization threshold the primary knock-on loses energy by making elastic collisions with other lattice atoms. When the energy of the primary event is less than E_d or the knock-on energy has been degraded below E_d, it is dissipated in the lattice as vibrational energy. Such local heating may be important in processes of radiation annealing. According to Kinchin and Pease (1955) the total number of atoms n_t displaced by a flux ϕ is given by:

$$n_t = N\sigma_s\phi(2 - L_e/E_{max})\frac{L_e}{4E_d}. \qquad (3)$$

Estimates of the damage for the oxides based upon these ideas indicate that for a given dose the crystals of highest density are damaged most, mainly as a result of the increased damage on the cation sub-lattice. Comparison of n_t (calculated using eqn. (3)) with the F^+ centre concentration (Henderson and King 1966) suggests that the calculations are in error by a factor of at least 10^2. This is not surprising since the effects of vacancy–interstitial recombination, thermal and displacement spikes, focusing and channelling, all of which reduce the amount of damage, are neglected. The importance of these effects in the oxides has recently been discussed by Wilks (1967) and will not further be discussed here.

During neutron irradiation both anions and cations are displaced, in relative concentrations proportional to their scattering cross section. Consequently, the defect structure of neutron irradiated oxides is likely to be much more complicated than that of the alkali halides damaged by ionizing radiation. In addition, we noted earlier that although the alkaline earth oxides have the rock salt structure, there are possibilities for many new defects merely because of the divalence of the host ions. Progress in understanding the defect structure of the oxides has been slow because techniques such as polarized luminescence and defect reorientation by polarized bleaching, so useful in studies of the alkali halides, have been unsuccessful in the alkaline earth oxides. For this reason, paramagnetic resonance has until recently been the most important technique in helping to describe the nature of radiation-induced defects in these insulators. How some of the defects (discovered by paramagnetic resonance) contribute to the coloration is yet to be determined.

§ 2. Principles of Magnetic Resonance and Optical Spectroscopy

Spectroscopy is concerned with the absorption or emission of electromagnetic radiation by atoms or systems of atoms according to the Bohr frequency condition

$$h\nu_{21} = E_2 - E_1. \tag{4}$$

Equation (4) expresses the energy conservation in a radiative transition of frequency ν_{21} between states of energy E_1 and E_2. In electron spin resonance (ESR) the atoms or defects have partially filled electronic configurations, a spin degenerate ground state and an associated non-zero magnetic moment. This ground state degeneracy is removed by an applied magnetic field as a consequence of there being several possible orientations of the magnetic moment. Transitions between adjacent levels, governed by a selection rule $\Delta M_s = \pm 1$, may be excited by radiation in the microwave region (8–50 GHz).

The frequency of the electromagnetic radiation detected in optical spectroscopy is much higher than in ESR: for a wavelength of 3300 Å the frequency is 10^{15} Hz. Selection rules in this case are related to those derived for electric dipole transitions in atomic spectroscopy (Woodgate, 1970). For states of well defined parity a very strict selection rule, Δl-odd, is derivable from symmetry arguments about the parity of states. Furthermore the selection rules of l and M_l are $\Delta M_l = \pm 1$, $\Delta M_l = 0$ for π-polarization and $\Delta M_l = \pm 1$ for σ-polarization, π- and σ-polarization referring to the experimental arrangement having the electric vector parallel and perpendicular to the applied magnetic field respectively. These rules are not necessarily as stringent for atoms or defects in crystalline solids.

2.1. Basic features of electron spin resonance

ESR is limited to the study of defects with $S \geq \frac{1}{2}$. For $S = \frac{1}{2}$ there are two components of the electronic spin ($M_s = \pm\frac{1}{2}$) which have energies

$$E = \pm\tfrac{1}{2}g\beta B, \tag{5}$$

g being the spectroscopic splitting factor, β the Bohr magneton and B the magnetic field (see fig. 3 (a)). Transitions between these levels are induced by applying an alternating magnetic field B_1 of frequency ν_0. If applied perpendicular to the steady field B_0, transitions occur when

$$h\nu = g\beta B_0 \qquad (6)$$

h being Planck's constant. The frequency ν of the electromagnetic radiation falls in the microwave region (8–50 GHz). Commercial spectrometers frequently operate at X-band with $\nu \cong 9.5 \times 10^9$ Hz; for a free electron with $g = 2.00$ the magnetic field

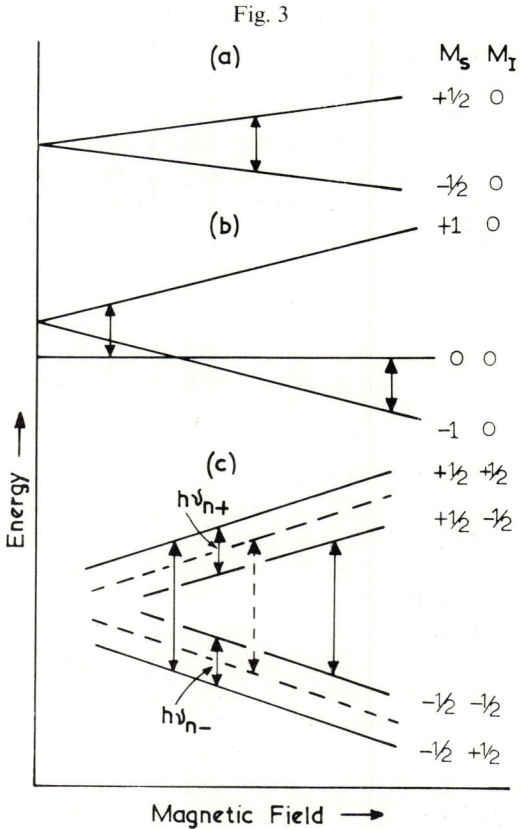

Fig. 3

Energy levels and transitions for a spin system as a function of applied magnetic field **B** and at constant microwave frequency. The examples given correspond to (a) $s = \frac{1}{2}$, $I = 0$, (b) $s = 1$, $I = 0$ and (c) $s = \frac{1}{2}$, $I = \frac{1}{2}$.

at resonance is about 3400 G. Because microwave sources can be tuned only over a narrow frequency range it is usual to search for resonances by keeping the frequency constant and varying the magnetic field.

The resonance can be detected only if there is a population difference between the two levels. Since the populations of the two levels are governed by Maxwell–Boltzmann statistics, there are always fewer electrons in the higher level. Thus there will be more transitions from the lower level, tending to equalize the populations. Except at very high microwave power level the populations are never equalized because after excitation to the higher level, the electrons relax from the higher state by phonon emission (so-called spin-lattice relaxation) in order to re-establish the Maxwell–Boltzmann distribution. Thus a steady absorption of microwave photons takes place. The net magnetic moment of the system is proportional to the population difference ($N_1 - N_2$) between the states,

$$N_1 - N_2 = N(e^x - 1)/(e^x + 1) \qquad (7)$$

where $N = N_1 + N_2$ is the total number of electrons in states 1 and 2 and $x = h\nu/kT$. Since for temperatures greater than ~ 4 K, $h\nu \ll kT$ we can expand the exponential to give

$$N_1 - N_2 = \frac{N}{2}\left(\frac{h\nu}{kT}\right) \qquad (8)$$

It follows that the magnetization and thus the sensitivity of the system increases with decreasing temperature and increasing microwave frequency. Although most spectrometers, commercial and custom-built, operate in the X-band range of microwave frequencies, there is obvious advantage to be obtained by working in the Ka band (~ 20 GHz) and Ku band (35 GHz). A second reason why low temperatures are often essential is that the spin-lattice relaxation time is longer at low temperatures. This has the effect of reducing considerably the width of the absorption line at low temperatures. The simple doublet state splitting, which produces a single line spectrum, is rarely to be met with in studying defect structure. It becomes necessary to consider all possible interactions between the electron spin and its environment. The simplest of such interactions is the coupling between the spin and angular momenta. This may arise from the electron's intrinsic orbital angular momentum (i.e. when $L \neq 0$), or when angular momentum is generated in the ground state through overlap of the electron wave function onto the orbitals of neighbouring atoms. Such wavefunction overlap causes fractional admixture of the ion-core functions into the ground state wavefunction. Spin-orbit coupling on the neighbouring ions is then manifest in second-order perturbation theory as a shift in the g-value for resonance absorption relative to that for free electrons. In addition the g-value, especially in low symmetry crystals, will depend upon the orientation relative to the applied magnetic field.

In some cases we are concerned with a system of several electrons with total spin $S > \frac{1}{2}$. The internal electric field of the crystal may then produce a splitting of the levels in the absence of a magnetic field. An example of the splitting of a spin triplet state with $S = 1$ is shown in fig. 3 (b), where the fine structure splitting leads to the $m_s = \pm 1$ and $m_s = 0$ levels being non-degenerate when $B_0 = 0$. The applied magnetic field completely removes the degeneracy and the levels split as shown in the diagram. Since the spin quantum numbers in allowed magnetic dipole transitions obey the selection rule $\Delta m_s = \pm 1$, we observe two transitions in the presence of microwave radiation of the correct frequency.

Finally we consider the interaction of the electron spin with the nuclear magnetic moments of neighbouring atoms.* This gives rise to hyperfine structure in the spectrum provided that the nuclear spin of the surrounding atoms is non-zero.

---

* In the special case where the defect on which the unpaired electron is located is an impurity atom, an important hyperfine interaction with the impurity nucleus may exist.

Figure 3 (c) shows the interaction when the $S = \frac{1}{2}$ spin doublet is coupled to a nuclear spin $I = \frac{1}{2}$. In a magnetic field, there are now four levels characterized by different values of the quantum numbers m_s and m_I. For allowed transitions we have to couple the additional selection rule $\Delta m_I = 0$ to the usual selection rule for magnetic dipole transitions.

The hyperfine interactions of defects are often exceedingly complex since they may involve the electron spins interacting with many nuclei. Consequently the electron spin resonance spectrum may consist of many overlapping lines, giving the appearance of a single broad, structureless line. In this case the hyperfine splittings are measured directly using electron nuclear double resonance (ENDOR). To illustrate ENDOR, we consider a system with total electron and nuclear spins $S = I = \frac{1}{2}$. If the electron spin resonance transitions have long spin-lattice relaxation times, the probability of exciting transitions from the upper to lower level is smaller than that for the reverse transitions. Consequently if sufficient microwave power is applied to the sample while the magnetic field is held at the resonant value for the electron spin system, a state may be reached in which the populations of both levels are equal. When this is so absorption of microwave power ceases and the transition is saturated. Now apply an intense radio frequency pulse and adjust the frequency until at some frequency ν_n, the transitions in fig. 3 (c) between the $m_I = +\frac{1}{2}$ and $m_I = -\frac{1}{2}$ levels are excited. At this point a slight depopulation of the $(\frac{1}{2}, \frac{1}{2})$ level results, removing the saturation condition, so that the electron spin resonance signal between the $(-\frac{1}{2}, \frac{1}{2})$ and $(\frac{1}{2}, \frac{1}{2})$ levels will rise again.

2.1.1. *The spin Hamiltonian*

The interpretation of an ESR or ENDOR spectrum may be quite complex due to the several electrostatic interactions which the unpaired electron must undergo in a crystalline environment. These interactions are described by the Hamiltonian,

$$\mathcal{H} = \mathcal{H}_C + \mathcal{H}_{LS} + \mathcal{H}_{SS} + \mathcal{H}_{ZE} + \mathcal{H}_{HF} + \mathcal{H}_Q + \mathcal{H}_{NZ}, \tag{9}$$

in which the terms are arranged in order of decreasing magnitude. $\mathcal{H}_C$ represents the sum of interactions of the electrons with both the nuclear charge and with each other including the crystalline electric fields. The different configurations resulting from this interaction are separated by energies of 3–5 eV (~ 10–$30{,}000$ cm^{-1}). $\mathcal{H}_{LS}$ is the relativistic spin-orbit coupling interaction which introduces splittings of order 10–1000 cm^{-1}. The mutual magnetic interaction between electron spins is represented by $\mathcal{H}_{SS}$: although not usually important in defects it should be considered for transition metal ions since in combination with $\mathcal{H}_{LS}$ it removes spin degeneracy in the absence of an externally applied magnetic field. Such splittings are usually less than 1 cm^{-1}. $\mathcal{H}_{ZE}$ and $\mathcal{H}_{ZN}$ are the Zeeman interactions between electronic and nuclear magnetic moments respectively, the latter being $\sim 1/2000$ smaller than $\mathcal{H}_{ZE}$. Consequently $\mathcal{H}_{ZN}$ is not usually measurable in ESR and resort to nuclear magnetic resonance (NMR) or ENDOR is required for an accurate determination. The presence of a hyperfine interaction, $\mathcal{H}_{HF}$, is a consequence of the mutual dipolar interaction between the nuclear and electronic magnetic moments in addition to the interaction of s-electrons and the nucleus. For nuclei with $I > \frac{1}{2}$ one observes an interaction of the nuclear electric quadrupole moment with any electric field gradients present at the nucleus. This is the origin of $\mathcal{H}_Q$, which can be measured directly using ENDOR techniques.

Fortunately the ESR data can be described in a fairly simple way without detailed knowledge of all the above effects. This is so mainly because the paramagnetic entity is present in a crystalline environment and consequently the crystalline electric field is invariant under the operations of the symmetry group to which the crystal belongs. Thus the Stark splittings of the energy levels in a particular crystal environment are calculable using the methods of group theory. An example of such splittings for the $3d^n$ configurations in weak or moderate crystal fields is given in fig. 4: no matter which ion has the particular $3d^n$ configuration these levels have the same order. Only the magnitude of Δ varies. In general Δ is large relative to the ESR quantum, spin-

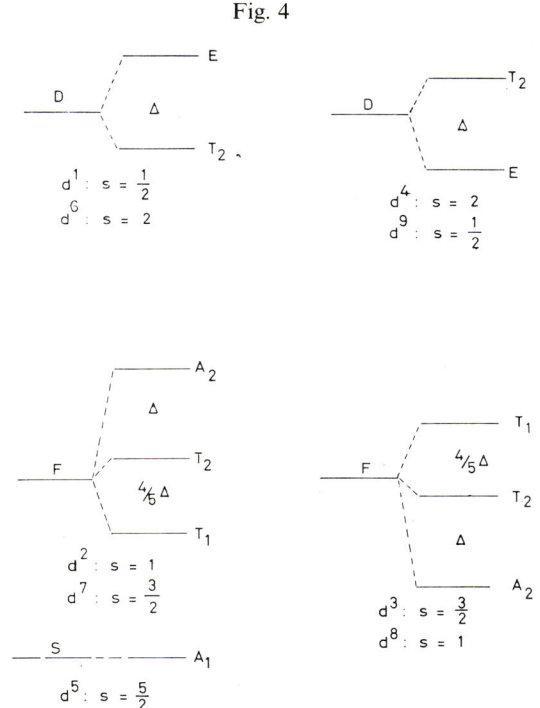

Fig. 4

Ground-state splittings of orbital levels for d^n ions in weak or in moderate crystal fields. Further splittings by spin–orbit coupling or other effects have been neglected. (After Low and Offenbacher 1965.)

orbit interaction, hyperfine interaction etc. Thus we can regard the lowest lying level in isolation. The magnetic behaviour of these levels can be described by an effective or fictitious spin, since we observe transitions between $2S + 1$ orientations of the ground state magnetic dipole in a magnetic field. A Hamiltonian can now be written which represents the magnetic behaviour of the ground state spin system only: since it contains polynomials in the electronic spin it is referred to as a *spin Hamiltonian*. Using this spin Hamiltonian the energy levels of the ground state in a magnetic field may be calculated using first and higher order perturbation theory (Abragam and Pryce 1950). The spin Hamiltonian is a convenient shorthand representation of the experimental results in which the measured splittings relating to g-values, fine-structure and hyperfine structure parameters etc. enter as energy terms. In general form we may write

$$\mathcal{H}_S = \sum \beta g_{ij}\vec{B}_i\vec{S}_j + D_{ij}\vec{S}_i\vec{S}_j + A_{ij}\vec{S}_i\vec{I}_j + P_{ij}\vec{I}_i\vec{I}_j - g_N\beta_N\vec{B}_j\vec{I}_j, \qquad (10)$$

where the indices i and j refer to the xyz co-ordinates and g_{ij}, D_{ij}, A_{ij}, P_{ij} are three-dimensional tensors of the second rank. There are numerous special forms of this spin Hamiltonian appropriate in cubic, tetragonal and lower symmetries crystals and these are fully described both in original papers and several textbooks (see e.g. Wertz and Bolton 1972). We will not deal with them further. However, since much of the following concerns defects which have $s = \frac{1}{2}$ and anisotropic hyperfine structure, this single instructional example will be discussed in some detail.

The ESR of most trapped electron and trapped hole centres in oxides is conveniently described by the spin-Hamiltonian

$$\mathcal{H} = \beta\vec{B}.\tilde{g}.\vec{s} + \vec{I}.\tilde{A}.\vec{s} - g_N\beta_N\vec{B}.\vec{I} \qquad (11)$$

in which β and β_N are electron and nuclear Bohr magnetons, s and I represent electron and nuclear spin operators, g_N is the nuclear g value for any neighbouring isotope, and the electronic g factor is in general a tensor quantity. The hyperfine interaction $\vec{I}.\tilde{A}.\vec{s}$ splits each electronic level into $2I + 1$ hyperfine levels, ESR transitions between the levels being governed by the selection rule $\Delta m_s = \pm 1$, $\Delta m_I = 0$.

The hyperfine tensor A may be written as the sum of a scalar a and a traceless tensor b. The scalar a is then referred to as the isotropic hyperfine interaction constant, and b describes any anisotropy in the hyperfine structure. Using first-order perturbation theory with the assumptions of high field ($g\beta B \gg a$) and small anisotropic interaction ($b_{ij} \ll a \pm g_N\beta_N B$), the eigenvalues of eqn. (11) are given by

$$E = m_s g\beta B - m_I g_N\beta_N B + m_s m_I(a + b_{z'z'}) \qquad (12)$$

where z' is the magnetic field direction. For the simplest case of a nucleus on an axis (z) of axial symmetry eqn. (12) reduces to

$$E = m_s g\beta B - m_I g_N\beta B + m_s m_I[a + b(3\cos^2\theta - 1)] \qquad (13)$$

where θ is the angle between B (the z' axis) and the symmetry axis z. Allowed EPR transitions are observed at magnetic field values.

$$h\nu/g\beta = B_0 = B + m_I[a + b(3\cos^2\theta - 1)] \qquad (14)$$

the constants a and b being expressed in magnetic field units. The isotropic or Fermi-Segrè interaction is related to the square of the wave function at the nucleus, $|\psi(0)|^2$, by

$$a = (8/3)\pi g\beta g_N\beta_N|\psi(0)|^2 \qquad (15)$$

where in CGS units $\beta = 0.92731 \times 10^{-20}$ erg/G, $\beta_N = 0.50504 \times 10^{-23}$ erg/G and $|\psi(0)|^2$ has units of cm^{-3} when a is measured in ergs. The conversion factors into alternative units in terms of the energy E in ergs are: $E = 10^6 h \times$ (MHz), $E = (hc) \times$ (cm^{-1}) and $E = (g\beta) \times$ (G).

The form of the wave function $\psi(0)$ used in the calculation of a depends on the approximate description of the ground state of the particular defect type. For example in a pseudopotential theory of the F$^+$ centre, the wave function is often written as

$$\psi = N\{\psi_F - \sum_i \psi_i\langle\psi_i|\psi_F\rangle\} \qquad (16)$$

where ψ_F is a smoothly varying, vacancy-centred envelope function, $\{\psi_i\}$ is the set of iron-core orbitals for the neighbours, N is a normalization factor, and ψ is orthogonalized to the ion-core wave functions ψ_i. Gourary and Adrian (1957) show that the substitution of ψ in eqn. (15) is equivalent to replacing $|\psi(0)|^2$ by $A_R|\psi_F(R)|^2$, where A_R is an 'amplification factor' for the nucleus at distance R from the vacancy centre. Holton and Blum (1962) have suggested that $A_R \propto (Z-2)^{3/3}$, Z being the atomic number of the nucleus concerned. The isotropic hyperfine interaction thus provides a measure of $|\psi_F(R)|^2$ and is related to the wave function overlap onto the neighbouring ion cores.

The anisotropic interaction, represented by a tensor b, depends upon the dipolar interaction between electronic and nuclear spins. In axial symmetry b may be written as

$$b = \tfrac{1}{2}g\beta g_N \beta_N \langle \psi | [3\cos^2 \alpha - 1]/|r|^3 |\psi \rangle \tag{17}$$

where r is the electron-nuclear separation and α is the angle between r and the symmetry axis z. Clearly the form of ψ indicates that several terms are involved in a proper calculation of b. Usually, the anisotropic interaction is determined mainly by overlap of ion-core p orbitals onto the ground-state wave function. The same matrix elements which determine this interaction are involved in the negative g shift of F centres relative to the free-electron g value.

Equation (14) describes the first order resonance fields when the hyperfine interaction involves only a single nucleus. However, around any defect or impurity ion site, there are many shells of near-neighbour cations and anions, all of which may contribute to the hyperfine interaction. The extent of this additional complexity depends largely upon the isotopic abundance of magnetic nuclei and the delocalization of the trapped electron. Thus, we have to sum eqn. (13) over all shells of interacting nuclei:

$$E = m_s g\beta B + \sum_n \{-M_I(n)g_N(n)\beta_N B + m_s M_I(n)[a(n) + b(n)(3\cos^2 \theta_n - 1)]\} \tag{13a}$$

Σ_n running over groups of equivalent nuclei. Thus eqn. (14) becomes

$$B_0 = B + \sum_n M_I(n)[a(n) + b(n)(3\cos^2 \theta_n - 1)] \tag{14a}$$

where the total nuclear spin quantum number $M_I(n)$ of the nth shell of equivalent nuclei is given by

$$M_I(n) = \sum_{i=1}^{x} m_I(i) = xI, xI-1, \ldots, -(xI-1), -xI \tag{18}$$

x being the number of equivalent nuclei in the nth group. The intensity of each hyperfine component depends upon the number of ways of compounding the total nuclear quantum number, as well as the abundance of the isotopic species.

Fortunately in the alkaline earth oxides the natural abundances of magnetic nuclei are low and these additional complications are largely ignored. However, the ENDOR technique discussed above, which would be necessary to unravel the many resonances implied by eqn. (14a), is still much used. From eqn. (12) it can be seen that there are two ENDOR transitions for each of the EPR transitions. Thus since the $m_I = \pm 1$ selection rule applies ENDOR transitions are observed at frequencies v_E given by,

$$v_E = h^{-1}\{-m_I g_N \beta_N B \pm \tfrac{1}{2} m_I (a + b_{z'z'})\} \qquad (14b)$$

from which we can see that the sum of the frequencies determines the nuclear g-value whereas the difference in frequencies gives the hyperfine interaction constants. When the nuclear species has magnetic spin $I > \tfrac{1}{2}$, a quadrupolar term enters explicitly into eqn. (19) and as a consequence may be measured directly in an ENDOR experiment. In interpreting the ENDOR spectrum it should be remembered that the linewidths are very narrow, being typical of NMR transitions rather than ESR transitions. Thus the hyperfine constants are determined with greater precision in the double resonance experiment, even when many overlapping lines are present. Furthermore the ENDOR technique permits a direct measure of the nuclear g-value, so identifying the interacting nucleus, in addition to the nuclear electric quadrupole interaction.

2.2. *Optical spectroscopy of solids*

The optical properties of solids are determined by the way in which the electrons in the material can respond to radiation. Optical techniques have been applied extremely successfully to the study of defects in ionic crystals, and to a lesser extent in semiconductors. Free electrons in metals cause almost total reflection of photons of all energies, and hence studies are restricted to reflectivity measurements. We confine this discussion to phenomena associated with absorption at energy levels in the band gaps of insulators and semiconductors.

Most insulators are transparent to visible light because a large energy gap exists between the valence and conduction bands. Impurity atoms and intrinsic lattice defects with energy levels in the band gap may cause selective absorption of some components of the visible spectrum: the crystals are then coloured. For example both KCl and MgO are transparent ionic crystals. After X-irradiation they exhibit in broad absorption bands in the red region of the spectrum and are characteristically blue in colour due to the presence of defects. The elemental semiconductors, Si and Ge, exhibit a metallic lustre when viewed in ordinary light. This is because photons with energy greater than about 1 eV excite electrons from states at the top of the valence band into the conduction band, with a consequent absorption of the incident photon. At shorter wave lengths especially in the infra-red these semiconductors become transparent.

Optical absorption and emission measurements are usually made at wave lengths between 1850 Å and 30,000 Å, in which range most commercial monochromators operate. The optical experiments give information about both the defect concentration and the nature of the electronic states involved in the transitions. It is usual in optical absorption experiments to measure the increased absorption of the crystal due to defects as a function of the wave length of light. The absorption coefficient, μ, is defined in terms of the radiation intensity, I, and the fractional intensity decrease $-dI$ per dt thickness by

$$dI = -\mu I \, dt.$$

Simple integration yields,

$$\mu = \frac{2\cdot 303}{t} \log_{10}\left(\frac{I_0}{I}\right). \qquad (19)$$

Furthermore the integrated absorption coefficient is related to the oscillator strength, f, of the transition since

$$\int \mu(E)\, dE = \frac{2\pi^2 e^2 h}{nmc^2}\left(\frac{E_0}{E_{\text{eff}}}\right)^2 Nf$$

where N is the volume concentration of defects, E_0 is the average electric field in the medium, E_{eff} is the effective field at the defect, and the other symbols have their usual significance. This equation was first solved by Smakula to give,

$$Nf = \text{const} \times \frac{10^{17} n}{(n^2+2)^2} \alpha_m W(T) \qquad (20)$$

in which n is the refractive index, α_m the band peak absorption coefficient, and $W(T)$ is the full band width at half the peak height. The constant is 1·29 for a Lorentzian band shape and 0·87 for a Gaussian band shape. Thus the concentration of centres can be computed from the absorption coefficient, if the oscillator strength of the transition is known.

The familiar broad bands of solid-state spectroscopy are a consequence of the coupling between the electronic states of the defects and the vibrational excitations of the lattice. Thus in an optical transition, absorption occurs over a range of energies corresponding to the vibrational energy levels associated with each of the electronic levels involved in the transition.

A more complete understanding of the electron–phonon interaction becomes possible if the band has a sharp structure associated with it when measured at low temperatures. The presence of narrow lines in vibrationally broadened optical transitions is quite common, and examples are to be found in atomic, solid-state and molecular spectroscopy. Delbecq and Pringsheim (1953) first reported fine structure attendant upon colour-centre bands in LiF: the structure consisted of a sharp line on the long wave length edge of the band and broader peaks extending to shorter wave lengths. These authors did not comment upon the significance of their observations. Wertz et al. (1963) reported similar observations in magnesium oxide after neutron irradiation without attempting to interpret the structure associated with any of the observed bands. Fitchen et al. (1963) showed that the phenomena observed by Delbecq and Pringsheim (1953) occurred in other alkali halides and interpreted the narrowest line as an optical transition in which no energy is transferred to the lattice phonons, i.e. a zero-phonon line. The succeeding peaks then correspond to the emission of one or more phonons of sharply defined frequencies.

Many authors have now utilized the zero-phonon line in attempts to probe directly the structure of defects in solids. This is because the existence of sharp lines, typically the widths are in the range 1–50 cm^{-1}, makes possible many experiments which are not possible using broad bands. The most obvious experiments are those involving the application of external perturbations; uniaxial stress, high magnetic fields (Zeeman effect) and high electric fields have all been used, although the second has been most useful. The theory of these transitions and of stress effects have been reviewed by numerous authors (Hughes 1966, Fitchen 1968, Lanzl et al. 1966); in the next section we discuss only those qualitative features of the theory necessary for understanding the phenomena.

2.2.1. *The configurational coordinate model*

The concepts of the zero-phonon and phonon-assisted transitions together with other aspects of optical absorption and emission spectra of defects in solids, are discussed most simply in terms of the configurational coordinate model. The physical basis for this model is based upon the adiabatic Born–Oppenheimer approximation used in the treatment of molecules (Lax 1952, Markham 1959). An exact calculation of the optical line shape for a colour-centre transition requires a knowledge of the electronic wave functions at the centre, the perturbed vibrational states of the lattice, and the form of the electron–phonon interaction at the defect. Lacking such knowledge, most models emphasize the qualitative features introduced by the electron–phonon interaction.

In the adiabatic approximation the eigenvalues of the electron moving under the influence of the oscillating field of force of the vibrating atoms of the crystal are calculated by treating the energy of the nuclei as a perturbation. The Schrödinger equation in this dynamic case is:

$$[\mathcal{H} - T - V(Q)]\Phi(r, Q) = E\Phi(r, Q), \tag{21}$$

where T and $V(Q)$ are respectively the kinetic and potential energies of the oscillator respectively, $\mathcal{H}$ is the Hamiltonian for the unperturbed electronic system and $\Phi(r, Q)$ is the total wave function of the system which varies only as a function of the electron, r, and nuclear coordinates, Q. The potential energy $V(Q)$ of the oscillator when expanded about the equilibrium position ($Q = 0$) is then $\frac{1}{2}\hbar\omega Q^2$, where Q is a dimensionless coordinate defined in terms of X, the coordinate relative to the equilibrium position, by $X = (h/M\omega)^{1/2}Q$. A central feature of the model is the assumption of product wave functions for the coupled defect of the form:

$$\begin{aligned}\text{ground state:} \quad & \Phi_{g,n} = \psi_g(r, Q)\chi_{g,n}(Q), \\ \text{excited state:} \quad & \Phi_{e,m} = \psi_e(r, Q)\chi'_{e,m}(Q),\end{aligned} \tag{22}$$

where n, m are vibrational quantum numbers and $\chi'_{e,m}(Q)$ is a harmonic oscillator wave function the centre of which is displaced relative to $\chi_{g,n}(Q)$ by an amount Δ. The Condon approximation assumes that the electronic wave functions $\psi(r, Q)$ vary only slowly with Q so that on substitution in (21) we obtain:

$$[W - T - \tfrac{1}{2}\hbar\omega Q^2]\chi(Q) = E\chi(Q), \tag{23}$$

where W is the eigenvalue of the unperturbed Hamiltonian. It is assumed that the vibrational frequency in the excited state is unchanged from the ground state. The quantum mechanical eigenvalues for the coupled system are then $E_g = (n + \frac{1}{2})\hbar\omega$ and $E_e = E_0 + (m + \frac{1}{2})\hbar\omega$. Thus transitions between the vibrational states m and n occur at energies $E_0 + (m + \frac{1}{2})\hbar\omega - (n + 1)\hbar\omega$, and the zero-phonon line occurs when $m = n$, at energy E_0. These transitions are shown on fig. 5, in which the values of E_g and E_e are plotted as a function of the nuclear coordinates Q, assuming the vibrations to be harmonic. Since linear coupling is assumed the shape of the parabolas are identical. These configurational coordinate curves can be used to illustrate most phenomena observed in optical experiments. At temperatures close to 0°K, only the $n = 0$ vibrational level is occupied, and thus transitions will occur only from this level. Since the time taken for the atoms around the defect to relax to the excited state configuration is long compared with the electronic excitation period, the transitions are represented by vertical transitions. Such transitions can, however,

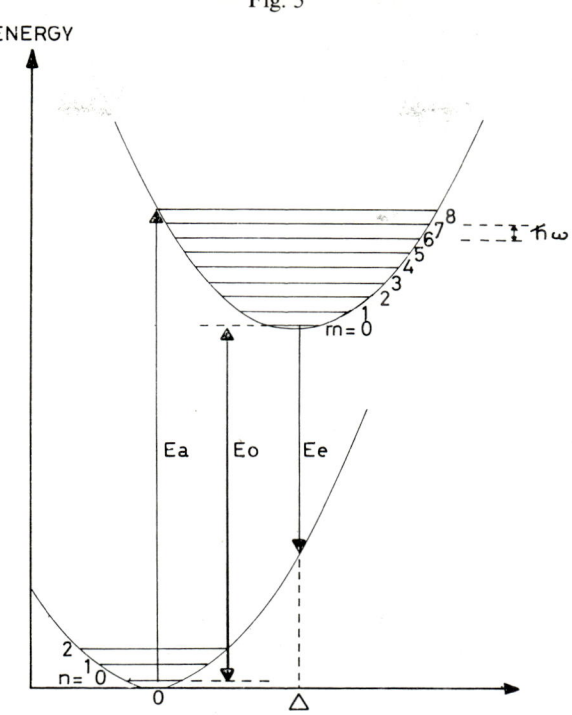

Fig. 5

Configurational coordinate curves for a defect involving a mean excitation of eight phonons in the excited state. For some defects (e.g. F^+ centre in MgO) the excited state relaxation is much greater, typically with $S > 20$.

take place with varying probability from any position consistent with the spatial extent of the $n = 0$ wave function, and thus the values of $E_0 + (m + \frac{1}{2})\hbar\omega$ define the width of the band at 0°K. At higher temperatures, levels other than $n = 0$ are occupied, and the transition energies defined above apply; the band is correspondingly broader. It is noted that once the electron has been excited into the upper level, the system can relax by phonon emission to the $m = 0$ level of the electronic excited state. Radiation may then occur as the electron returns to the ground state with energies given by $E_0 - (n + \frac{1}{2})\hbar\omega - (m + \frac{1}{2})\hbar\omega$; thus the peak in the emission band is shifted relative to the absorption band (Stokes shift).

The zero-phonon transition occurs midway between the absorption and emission bands. This transition will only be observed against the background of the broad band if it occurs with high probability; i.e. if there is a large overlap between the wave functions in the ground and excited state. The probability for an electric dipole transition between the states $\Phi_{g,n}$ and $\Phi_{e,m}$ is proportional to the square of the matrix elements connecting these states, i.e. $P_{ge} = |\langle\Phi_{e,m}|\mathbf{r}|\Phi_{g,n}\rangle|^2$. In the Condon approximation, substituting the wave functions (22) the probability becomes

$$P_{ge} = |\langle\psi_g|\mathbf{r}|\psi_e\rangle\langle\chi_m|\chi_n\rangle|^2. \tag{24}$$

The term $\langle\psi_g|\mathbf{r}|\psi_e\rangle^2$ is a constant, viz. the oscillator strength of the transition. Thus the transition probability is just the overlap integral for harmonic oscillator wave functions of the vibrational n, and m levels. This overlap integral between the displaced harmonic oscillator wave functions $\chi(Q)$ and $\chi'(Q)$ has been evaluated in numerous papers (Hughes 1966, Fitchen 1968, Lanzl et al. 1966, Markham 1959). For the low temperature case, when $n = 0$:

$$P_{0m} = \frac{e^{-s}S^m}{m!}, \qquad (25)$$

where S is the coupling or Huang Rhys factor defined in terms of Δ, the shift of the equilibrium coordinate value in the excited state relative to the ground state by $S = \frac{1}{2}\Delta^2$. Equation (25) defines the shape of the absorption band at low temperatures. This model implies strong coupling of a single mode to the electronic states at the defect. In practice this treatment can easily be extended to the opposite extreme more often met with in practice, of weak linear coupling to many modes (Hughes 1966, 1967). It is also possible to include the effects of temperature, since above $0°K$ the levels with $n > 0$ becomes thermally populated. In this case the zero-phonon probability becomes:

$$P_{00} = \exp(-S \coth \hbar\omega/2kT). \qquad (26)$$

This function decreases with increasing temperature, until it reaches a limiting value of unity at $T = \hbar\omega/4K$. For magnesium oxide $\omega \approx 300\,\text{cm}^{-1}$; hence this critical temperature, below which the zero-phonon line probability does not increase, is about $110°K$. Lower temperatures are, however, usually required since the linewidth is also temperature dependent as a result of quadratic coupling to the lattice modes. Thus, in magnesium oxide, temperatures in the liquid nitrogen range are imperative, and temperatures in the liquid helium range have proved most useful.

Examples of the shape function defined by eqn. (25) for chosen values of S are given in fig. 6. It is apparent that for very small values of S, most of the intensity is in the zero-phonon transition. When S is large, the probability of observing the zero-phonon line is negligible and the maximum value of P_{m0} is at $m = S$. The shape then approximates to a Gaussian distribution. This is intuitively obvious, since the larger the value of S, the larger is the value of Δ and the smaller the overlap in the lowest occupied level of the harmonic oscillator wave functions. The appearance of the Gaussian band shape in the limit of large S values shows that the configurational coordinate model discussed here in terms of the zero-phonon transition includes the theory of the broad structureless bands implicitly. Lax (1952) has discussed the properties of these bands in terms of moments of the band shape. For $S > 2$ the full width at half-height is related to the second moment at $T°K$ by:

$$M_2(T) = H(T)^2/5.6 = S(\hbar\omega)^2 \coth \hbar\omega/2kT,$$

so that the zero-temperature half-width $H(0)$ is related to S by:

$$S = [H(0)/2 \cdot 36\hbar\omega]^2. \qquad (27)$$

We will make use of these relationships in discussing the temperature dependence of the F^+ band in § 5.2.

Let us return briefly to the zero-phonon transition, which relative to the whole band has probability e^{-s}. That the transition is observed at all is due to the transition probability increasing exponentially with decreasing S values and to the line not being broadened by the finite frequency spread of the normal lattice models. This

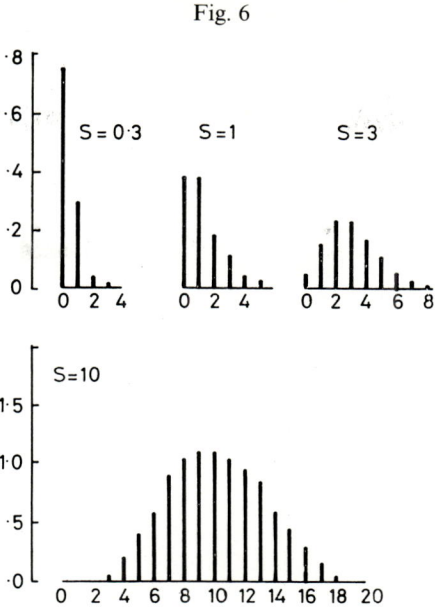

Fig. 6

The shape functions for absorption bands with different values of S. In the limit of large S the band peak occurs at $A = m$.

broadens only the phonon-assisted peaks so that at low temperatures the width of the zero-phonon line should in a perfect crystal at 0°K be determined only by lifetime broadening. Inhomogeneous broadening by random lattice strains results in widths of order $1-50$ cm^{-1}. Even with such widths, splittings of the lines by readily available strain amplitudes may be achieved in numerous cases. It is proposed here to outline the general principles of the theory of stress effects worked out independently by Kaplianskii (1964) and by Runciman (1965); the studies of zero-phonon lines and the application of uniaxial stress to zero-phonon lines in magnesium oxide and calcium oxide are discussed later.

2.2.2. Uniaxial stress effects

Studies of the effects of magnetic and electric fields on absorption and emission spectra have been common features of atomic and molecular spectroscopy for many years. In solids one may also apply uniaxial stress, and this has proved a most useful tool in probing the structure of defects in insulators. The principles involved are quite general for all three perturbations, since each may change the electronic energy levels under consideration and so remove certain forms of degeneracy associated with the energy levels. Since the energy levels are changed, the spectroscopic line may be shifted and in some cases split. The electronic states and symmetry properties of the defect may be determined from the splitting pattern. This discussion considers only the effects of uniaxial stress, since this has proved the most useful experimental technique.

The usual experimental procedure for cubic crystals is to apply stresses along $\langle 100 \rangle$, $\langle 110 \rangle$ and $\langle 111 \rangle$ crystal symmetry axes, and to examine the spectra with light polarized with the electric vector parallel (π) and perpendicular (σ) to the stress direction. Under ⟨110⟩ stress, the crystal becomes biaxial, and the number of components under σ illumination depends upon the direction of viewing (usually we choose the ⟨001⟩ and ⟨110⟩ axes as propagation directions), whereas for ⟨110⟩, and ⟨111⟩ stress the direction of viewing is of no consequence. The quantities to be determined experimentally are: the number of components for a given stress direction, the energy shift of each component, and the intensity and polarization of the components.

The types of defect which may occur in cubic crystals may be divided into three classes according to the types of degeneracy involved. In the simplest case, the splitting arises only because the defect has non-degenerate electronic states and is anisotropic. Hence different equivalent directions in the unstressed crystal become inequivalent when stress is applied. This type of degeneracy is called orientational degeneracy, and there are seven basic types of anisotropic centres in cubic crystals. Schematic models of these centres and the point groups comprising each type are shown in fig. 7. In unpolarized light when stress is applied, a single line of equal intensity will be seen for each of the equivalent orientations of the centre, since all equivalent orientations are equally populated if the mechanism of defect production is random. If the point group of the perfect lattice has S elements and that of the defect s elements, there will be S/s lines or equivalent orientations. The number and

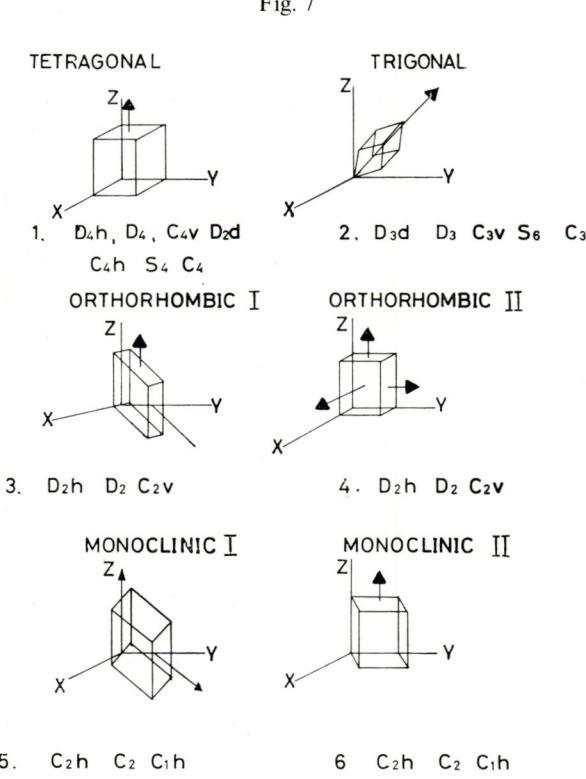

Fig. 7

The basic types of anisotropic centre and corresponding point groups in cubic crystals. The most general case, that of triclinic symmetry is omitted. (After Hughes 1966).

intensities of components and the magnitudes of the splittings for the seven types of anisotropy have been given previously by Kaplianskii (1964).

The second class of defects involves those having octahedral or tetrahedral symmetry, and an applied stress removes the electronic degeneracy of the states of a single defect. There are only six possible types of spectra within this category, corresponding to the various allowed transitions between states transforming like bases for irreducible representations of the cubic groups. The number of components may be found by noting the way in which the representation of a state belonging to a particular irreducible representation splits up into irreducible representations of that sub-group to which the defect in the deformed crystals belongs. In practice, however, it is frequently easier to perform the analysis by inspection of the angular form of the wave function. The electric dipole transition associated with an F centre is an especially easy case to consider, since the transition occurs between s and p-like states. Thus the angular form of the wave functions forms bases for the irreducible representations $^2A_1$ (ground state) and $^2T_1$ (excited state). Under a [100] stress the crystal has tetragonal symmetry, and the $^2A_1$ state is unaffected. The basis function of T_1 will break up under [100] stress into a doublet (E) and a singlet (A), so that two components are observed.

The final class of defects possess both orientational and electronic degeneracy, and an applied uniaxial stress may remove both types. Only defects with tetragonal or trigonal symmetry having doubly degenerate electronic states (E states) are involved: the number and intensity of the components have been calculated by Runciman (1965), and Hughes and Runciman (1967) have determined the stress shifts in terms of a number of arbitrary parameters.

2.3. Experimental considerations

No matter with what type of spectroscopy we are concerned, excitations of transitions between the various energy levels follow as a consequence of the sample being subjected to electromagnetic radiation. In most cases it is the electric field component of the electromagnetic wave which interacts with the sample so inducing electric dipole transitions. For example, electromagnetic radiation in the near ultra-violet or visible region will excite transitions between the electronic states of atoms in the solid. By comparison molecular rotations or vibrations also create oscillating electric dipoles with frequencies in the microwave or infra-red regions respectively.

The results are usually presented graphically as plots of functions which represent the absorbed (or emitted) intensity versus the photon energy. The absorbed intensity may be measured by the absorption coefficient as discussed in § 2.2. However in most commercial spectrophotometers the recorder slidewire is calibrated in terms of transmission, $T = I/I_0$, (often expressed as a percentage) or the optical density, OD (absorbance). These experimental observables are related to the absorption coefficient, μ, by

$$OD = \log_{10}(1/T) = \mu t/2 \cdot 303. \tag{28}$$

In luminescence studies most authors quote the emission in 'arbitrary units', an indication of the difficulty in accurately calibrating the detection system. The easiest method of representing the photon energy is in terms of the wave length λ, the units of which are the nanometre (nm), the micron (μm) or the Angström (Å). More convenient, since it is directly proportional to the photon energy is the wave number

$\bar{\nu} = 1/\lambda$, which has units of cm^{-1}. Spectroscopists who measure broad band spectra may plot results directly as a function of the photon energy ($E = h\nu = hc\bar{\nu}$) measured in electron volts.

In fig. 8 is shown a simple single beam spectrometer. Typically in absorption, the straight-through configuration, we use a continuous source such as tungsten, or

Fig. 8

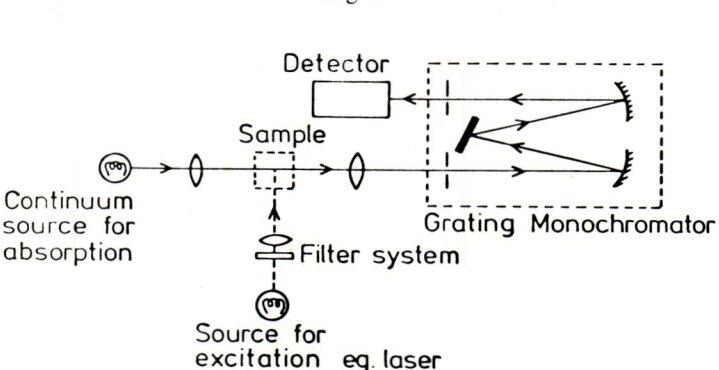

High resolution single beam spectrophotometer (after Hughes 1976, Defects and their Structure in Non-Metallic Solids (Ed. Henderson and Hughes), Plenum Press, N.Y.).

hydrogen lamp. The detection system usually comprises an electron multiplier phototube, electrometer amplifier and pen recorder. A variety of phototubes may be used depending upon the wave length range of interest. The recently developed GaAs tubes are extremely useful in view of the essentially flat response over the range 270–900 nm. The sensitivity is likely to be of order 10^{15}–10^{17} centres cm^{-3} in absorption and 10^{11} centres cm^{-3} in emission. Such a system although of low sensitivity is most likely to be used in high resolution measurements. The optimum resolution is determined by the monochromator. A modern grating monochromator of ca 1 m focal length will have a first order dispersion of 1 nm mm^{-1} and be capable of a resolution of 0·01 Å. Thus this method has great value in perturbation spectroscopy of sharp zero-phonon lines. In low resolution studies one might use a prism or a grating with only a few hundred ruled lines per mm.

The sensitivity may be greatly enhanced by converting the single beam' system into a double beam instrument.

A conventional double beam spectrophotometer for absorption studies is shown in fig. 9 (a): this is the principle of a number of commercial instruments. The beam, after passage through the grating monochromator, is divided into sample and reference channels by a beam splitting chopper. After recombination at the phototube, phase sensitive detection and amplification is used to obtain a signal suitable for output to a pen recorder. Note that by chopping the light beam at a pre-selected frequency permits narrow band amplification of the detected signal. Thus any noise components in the signal are limited to a narrow band centred at the chopping frequency. The signal obtained has the normal absorption bandshape. However in many absorption and emission investigations the results are presented in differential form. This type of differential spectroscopy is useful in studying the effects of stress, magnetic field and other external perturbations on defect energy levels. For example in magnetic circular dichroism we apply a static magnetic field to remove spin degeneracy of the states and optical transitions in the static field may then be circularly or linearly polarized. Very detailed information about spin-

Fig. 9

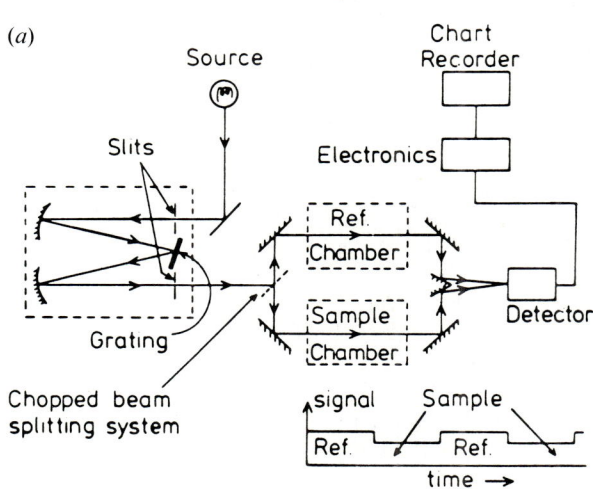

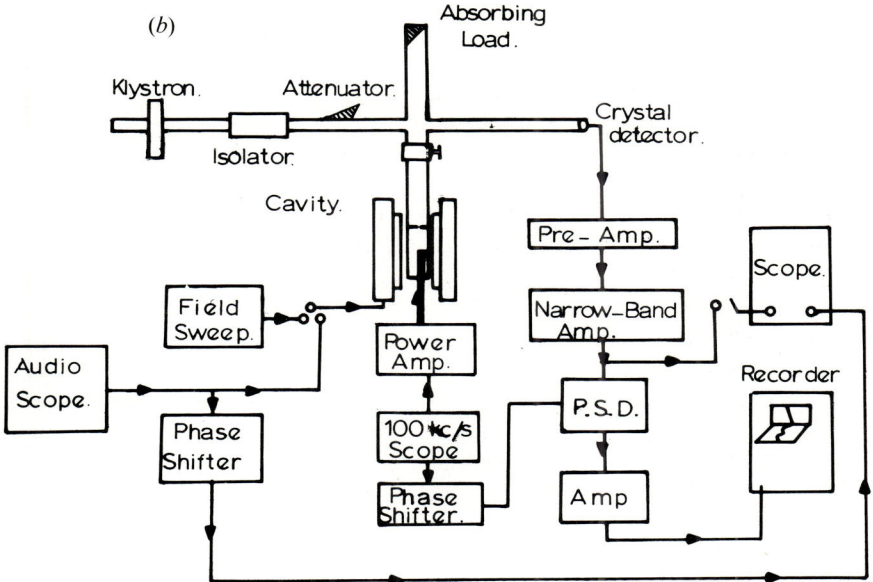

Comparison of (a) double beam optical spectrophotometer (after Hughes 1976, Defects and their Structure in Non-Metallic Solids (Ed. Henderson and Hughes), Plenum, N.Y.) and (b) a conventional ESR spectrometer (after Wertz and Bolton 1972).

orbit couplings, Jahn–Teller effects and the electron-phonon interaction may be obtained by modulating the circular polarization (σ_+, σ_-) of the absorbed or emitted light. A simple descriptive account of this type of measurement is given by Merle d'Aubigné (1976): the basic principles and theory may be found in Henry, Schnatterly and Slichter (1965) and Henry and Slichter (1968).

In electron spin resonance spectroscopy, magnetic dipole transitions are induced by the oscillating magnetic field component of the electromagnetic radiation. The sample contained in a resonant cavity must be placed in an homogeneous applied magnetic field. The purpose of the resonant cavity is to enhance the energy density of the microwaves at the sample. The spectrometer shown in fig. 9 (b) is analogous to the double beam spectrometer in fig. 9 (a). To enhance the sensitivity the magnetic field is modulated at some convenient frequency: this plays an analogous role to the light chopper in the optical case. The continuous source of electromagnetic radiation is a klystron, the frequency of which is controlled to an accuracy of ~ 1 in 10^6 Hz by means of suitable electronics. The cavity system, including the waveguide arrangement to control and direct microwave power onto and away from the sample, plays much the same role as the optics in the spectrophotometers described earlier. There are many variations on this particular scheme: the reader is referred to the texts produced by Poole (1967) and Wilmshurst (1967).

§ 3. Optical and magnetic properties of impurity centres

Ions of the $3d^n$ series ('iron group') readily enter substitutional sites in the alkaline earth oxides. In fact, neither of the authors has yet been able to obtain a single crystal of MgO, CaO or SrO which did not show the ESR spectrum of one or more substitutional impurity ions. In doped MgO, some valence state of all iron-group elements except Sc^2 has been seen. A few of the 4d ions have also been observed in doped MgO. Some of the rare-earth ions may also be detected in CaO or SrO; only Er^{3+} has been detected in MgO. It is not necessary that the substituting cation be divalent; in MgO the trivalent ions V^{3+}, Cr^{3+}, Fe^{3+} and the tetravalent ions Ti^{4+} and Mn^{4+} can represent stable substituents. Some valence states may be altered by heat treatment. Unstable monovalent ions which may be formed by irradiation are Fe^{1+} and Co^{1+}. Less stable divalent ions formed by irradiation include V^{2+} and Cr^{2+}. Trivalent ions formed during irradiation include Fe^{3+} and Ni^{3+}. Analysis of MgO universally shows the presence of aluminium and silicon. The former is indicated in § 3.1 to be involved in some associated centres. The site of the silicon in these oxides is presently unknown. Evidence for well-defined centres involving specific anionic impurities will be given in § 4. Naturally, all charge deviations must be compensated; the varieties of locally or remotely compensated centres thus far found will be considered in § 3.1. The greatest intensities of iron-group ESR lines are associated with ions which are in essentially octahedral environment. 'Essentially octahedral' here implies that there is no other impurity or vacancy within a circumscribing sphere of radius equal to two unit-cell lengths. Otherwise the ESR spectrum would deviate markedly from the very simple behaviour which is commonly observed, e.g., isotropic lines. This is in marked contrast to other solids (such as semiconductors) in which there is usually a close association between a defect with an abnormal charge and the compensating defect. Departures from exact octahedral symmetry are usually ascribed to a distribution of strains in the crystal. Tetragonal distortions are usually the most significant in the alkaline earth oxides. Manifestations of such strain distributions are seen in the isotropic ESR lines of Fe^{2+} and

Ni^{2+}, which may have widths respectively of the order of 500 and of 50 G at 20°K. For each of these the effective spin of the lowest-lying level is 1. Hence each of these represents a set of three levels between which there would be two transitions of equal energy if exact octahedral symmetry obtained. For such ions as Cr^{3+}, the width is sometimes less than 0·5 G. However, close examination of the base of the (integrated) absorption line of Cr^{3+} reveals a pair of shoulders. These are indicative of the fact that in a slightly-distorted octahedral field, the $\frac{3}{2} \leftrightarrow \frac{1}{2}$ and $-\frac{1}{2} \leftrightarrow -\frac{3}{2}$ transitions occur at field values slightly different from that of the $\frac{1}{2} \leftrightarrow -\frac{1}{2}$ transition. In an exactly octahedral field, the resonant field values for the three transitions would coincide. Naturally, it is essential to be aware of such matters when using the intensity of a derivative line as a presumed measure of the concentration of some ionic component. In some cases one has hints that there is distortion present because a line may broaden markedly as the field is rotated away from a [100] direction in a (100) plane. In other cases, line widths may be a minimum when $B\|[111]$, implying the occurrence of a small range of tetragonal distortions. In contrast with the iron-group ions, the rare-earth ions are often found in sites of tetragonal symmetry. For the iron-group ions this commonly implies association with a positive ion vacancy; however, for the rare earths, one is largely ignorant of the source of the tetragonal distortion.

The reduced magnitude of the crystalline electric field in CaO, SrO and BaO as compared with MgO leads to a number of differences between the spectrum of a given ion in these hosts. Among these differences are:

(a) A shortened spin-lattice relaxation time in CaO or SrO due to a decreased separation of interacting excited states. Observation of Fe^{3+} or Ni^{2+} with line widths as narrow as those in MgO requires considerably lower temperatures.

(b) The g factors of Cr^{3+}, Co^{2+} and Ni^{2+} show a larger deviation from the free-electron value. This is attributed to a smaller value of Δ in the expression:

$$g = 2(m - n\lambda/\Delta).$$

(c) A reduced octahedral zero-eld splitting for S-state ions in CaO and SrO makes it possible to see the $+\frac{1}{2} \leftrightarrow -\frac{1}{2}$ transitions in powders. For Fe^3 in MgO, one observes only the turning points for this transition. While Gd^{3+} exhibits a very asymmetric line in MgO powders, it shows a symmetric and much narrower line in CaO, SrO and BaO powders. There is evidence of fine structure for Gd^{3+} even in powdered CaO; the fine structure lines are located much closer to the central line in the case of SrO. Tables 2 to 6, which give the spectral parameters for all transition metal and rare-earth ion impurities respectively, are extended and revised versions of those given by Low and Offenbacher (1965). Appropriate references, except where specifically noted are to be found in that paper.

3.1. Electron spin resonance spectra of transition metal ions

The spectroscopic states of the $3d^1$ to $3d^9$ transition metal ions are as follows:

D: d^1, d^4, d^6 and d^9,

F: d^2, d^3, d^7 and d^8,

S: d^5.

The partial removal of orbital degeneracy of the D and F-state ions upon their substitution for a host cation in octahedral symmetry leads to the distinctive splitting patterns given in fig. 4. Further splittings from spin–orbit coupling or from other effects have not been shown. The orbitally non-degenerate state of S-state ions is also shown. From the magnetic resonance point of view, the d^3 and the d^5 or d^8 configurations show immediate promise because the lowest level is orbitally non-degenerate. Especially for the d^3 and d^5 ions, one can be assured of making spin transitions in a magnetic field because there will be at least a two-fold Kramers degeneracy which can only be removed by the application of a magnetic field. One observes ESR transitions for d^8 ions also. However, for the remaining ions, one must look for additional mechanisms to remove the orbital degeneracy of the ground level. These will be considered in a brief discussion of the d^1, d^6, d^7 and d^9 cases. Thus far no ESR spectra of d^2 or d^4 ions have been observed in the octahedral oxides.

For the d^1 ion Ti^{3+}, the introduction of a tetragonal distortion splits both the T_2 and the E states of fig. 4; the former is split into an orbital singlet (B_2) and a doublet (E) fig. 10. In MgO, the tetragonal distortion arises from a nnn cation vacancy (i.e. a vacancy along an [001] direction). The lowest-lying state is the orbital

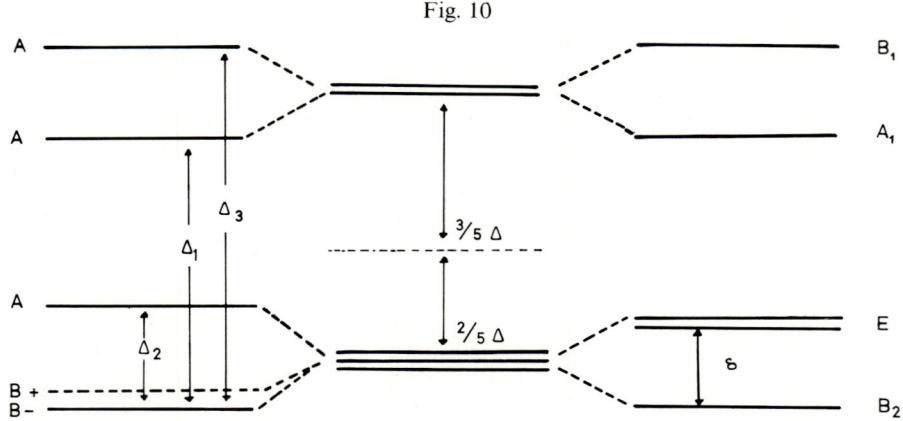

Fig. 10

Splittings of the energy levels of a d_1 ion in fields of tetragonal and orthorhombic symmetry.

singlet described by $2^{-1/2}(|2\rangle - |-2\rangle)$. Lying above this state by an energy δ are the degenerate states $|+1\rangle$ and $|-1\rangle$. For the magnetic field along the tetragonal axis, the spin–orbit coupling mixes some of the states $|\pm 1, \pm\frac{1}{2}\rangle$ and $|(|2\rangle + |-2\rangle)$, $\pm\frac{1}{2}\rangle$ with the $M_S = \mp\frac{1}{2}$ components of the ground state (Pake 1962). Evaluation of the diagonal elements of the operator $\mathbf{L}_z + 2\mathbf{S}_z$ with the modified ground-state wave function leads to the energies; from their difference one obtains the value of:

$$g_\| = 2 - \frac{8\lambda'}{\Delta} - \frac{\lambda'^2}{\delta^2}. \tag{29a}$$

When the magnetic field is perpendicular to the tetragonal axis, the matrix elements to be evaluated are those of the operator $\mathbf{L}_x + 2\mathbf{S}_x$. These lead to a 2×2 secular determinant, from which one obtains the energies:

$$W = \pm\left(1 - \frac{\lambda'}{\delta}\right)\beta H_x,$$

and hence:

$$g_\perp = 2 - 2\frac{\lambda'}{\delta}. \tag{29b}$$

ESR spectra due to Ti^{3+} have been observed in both MgO and CaO. A spectrum at X-band with $g_\parallel = 1\cdot9533$ and $g_\perp = 1\cdot8993$ in MgO was shown in later K-band measurements of the $^{47,49}Ti$ hyperfine structure to be due to Ti^{3+} in tetragonal symmetry sites (Wertz et al. 1964, Davies and Wertz, 1969). For $87\cdot2\%$ of Ti nuclei with $I = 0$ the ESR spectrum at arbitrary orientation consists of three lines of equal intensity due to equal population of centres with tetragonal axes parallel to the mutually orthogonal $\langle 100 \rangle$ axes. With $\vec{B}$ along the [001] crystal direction a line of single intensity ($g_\parallel$) is observed for centres having their distortion axis parallel to the field, together with a doubly intense line ($g_\perp$) due to centres with axes perpendicular to the field. Measurements at high amplifier gain show the $g_\parallel$ line to be attended by a well resolved, albeit weak, hyperfine structure due to the $7\cdot3\%$ abundant ^{47}Ti ($I = \frac{5}{2}$) and $5\cdot5\%$ abundant ^{49}Ti ($I = \frac{7}{2}$) nuclei. Figure 11 shows that observation of this hyperfine structure is hampered by serious interference from other impurity spectra. In the perpendicular orientation the hyperfine structure is broadened so that the lines are not so well resolved. Similar effects are observed in CaO.

These spectra may be fitted to eqn. (11) using the spin Hamiltonian parameters given in table 2. The crystal field splittings (fig. 10) are obtained from the expressions

Fig. 11

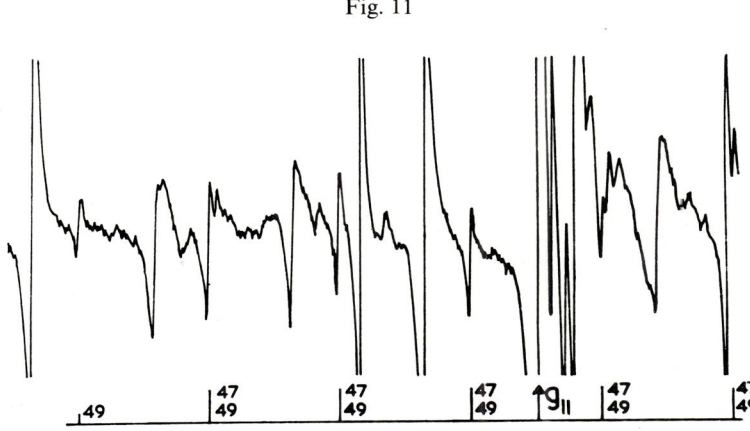

The hyperfine structure due to ^{47}Ti and ^{49}Ti in the ESR spectrum due to axial Ti^{3+} in MgO. The spectra were recorded at $77°K$ and $\nu = 17\cdot46$ GHz. Interference due to impurities (mainly Mn^{2+} and V^{2+}) is particularly notable on high field side of $g_\parallel$ (after Davies and Wertz 1969).

for $g_\parallel$ and $g_\perp$ using the free ion value of $\lambda' = 154$ cm^{-1}. Hence we find $\delta = 3000$ cm^{-1} from $g_\perp$ and $\Delta_1 \approx 30,000$ cm^{-1} from $g_\parallel$. For the corresponding centre in CaO $\delta \approx 4800$ cm^{-1} and $\Delta_1 \approx 22,000$ cm^{-1}. Note that since λ' is probably too large the magnitudes of δ and Δ_1 estimated in this way are probably too large. The origin of the tetragonal distortion is most likely a cation vacancy associated with the Ti^{3+} ion in a nearest neighbour site along a $\langle 100 \rangle$ axis (fig. 12 (a)). For this model Davies

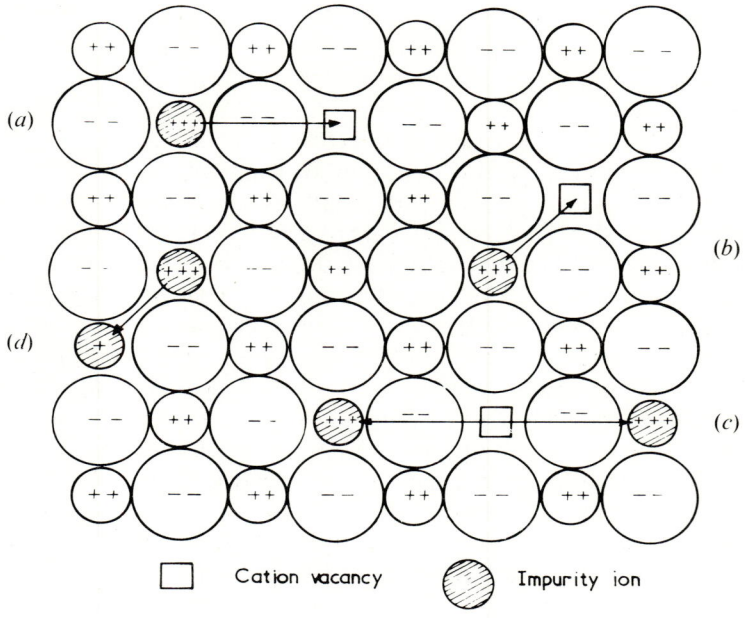

Fig. 12

Models of charge compensation mechanisms in oxide crystals.

and Wertz (1969) have used a point charge approach to predict the correct sign for δ: they find $\delta \approx 1700$ cm^{-1}.

In Ti-doped MgO an orthorhombic ESR spectrum is observed at temperatures below 20°K. The ^{47,49}Ti hyperfine structure is resolved with $\vec{B}$ parallel to the z-direction. Davies and Wertz (1970) ascribe the spectrum, the spin Hamiltonian parameters of which are given in table 2, to Ti^{3+} ions with nearest neighbour cation vacancies in the [110] direction (fig. 12). As shown in fig. 10 a crystal field of ortho-

Table 2. Spin Hamiltonian parameters for Ti^{3+} ions

Host	Symmetry	Temp (°K)	g-values	A values 10^{-4} cm^{-1}	Reference
MgO	tetragonal	77	$g_\parallel = 1.95304 \pm 0.0010$	$^{47,49}A_\parallel = 26.8 \pm 0.2$	Wertz et al. 1964
			$g_\perp = 1.89878 \pm 0.00010$	$^{47,49}A_\perp = 10.0 \pm 0.5$	Davies and Wertz 1969
CaO	,,	77	$g_\parallel = 1.9427 \pm 0.0001$	$^{47,49}A_\parallel = 27.4 \pm 0.1$	Davies and Wertz 1969
			$g_\perp = 1.9380 \pm 0.0001$	$^{47,49}A_\perp = 9.9 \pm 1.5$	
MgO	orthorhombic	20	$g_x = 1.8487 \pm 0.0007$		Davies and Wertz 1970
			$g_y = 1.9464 \pm 0.0007$		
			$g_z = 1.7670 \pm 0.0007$	$^{47,49}A_z \sim 12.5$	

rhombic symmetry splits the T_{2g} level of fig. 4 into three orbital singlets A, B_+ and B_-. ESR experiments do not distinguish between $B_+ = 2^{-1/2}(|yz\rangle + |xz\rangle)$ and $B_- = 2^{-1/2}(|yz\rangle - |xz\rangle)$. However both B_+ and B_- should lie lower than A, since the lobes of $A \approx |xy\rangle$ point directly at the vacancy. In terms of the splittings in fig. 10 the g-values for B_- are

$$g_x = 2 \cdot 0023 - \frac{2\lambda'}{\Delta_2} - \frac{6\lambda'}{\Delta_3}$$

$$g_y = 2 \cdot 0023 - \frac{2\lambda'}{\Delta_1} \quad (29c)$$

$$g_z = 2 \cdot 0023 - \frac{2\lambda'}{\delta}$$

where x, y, and z are taken to be parallel to [110], [1̄10] and [001] respectively. The relative choice of x and y is arbitrary. Thus δ is estimated from g_z to be ca 1300 cm^{-1}.

Taking g_y for B_- to correspond to the measured g_y gives $\Delta_1 = 5500$ cm^{-1}, a very large distortion of the octahedral field. Assuming Δ_3 to be of the same order of magnitude as $\Delta_1 \approx 30{,}000$ cm^{-1} in the tetragonal case we find $\Delta_2 \approx 3300$ cm^{-1} using g_x. Some support for these assignments are obtained from the hyperfine structure data. Expressions for A_z are derivable using the methods of Abragam and Pryce (1951): Davies and Wertz (1969) find that for B_+ or B_-

$$A_z = 2\delta_N \mu_B \mu_N \langle r^{-3} \rangle_{3d} \left\{ -\mathcal{K} + \frac{2}{7} - \frac{2\lambda}{\delta} + \frac{3\lambda}{7}\left(\frac{1}{\Delta_2} + \frac{1}{\Delta_1} - \frac{1}{\Delta_3}\right) \right\}. \quad (30)$$

Using values of $\langle r^{-3} \rangle_{3d}$ and the core polarization constant determined for the tetragonal centre gives $A_z = 12 \cdot 3 \times 10^{-4}$ cm^{-1}, in good agreement with the experimental value. Assuming the ground state to be A leads to an overestimate of A_z by a factor of two.

The d^3 ions are typified by Cr^{3+}, with $S = \frac{3}{2}$. ESR spectra of the isoelectronic ions V^{2+} and Mn^{4+} in octahedral symmetry are also seen in MgO. The first member of the series, Ti^{1+}, has recently been seen in octahedral symmetry in CaO (Tomlinson and Henderson 1968). Consideration of the spin–orbit coupling readily gives for the isotropic g factor:

$$g = 2 \cdot 0023 - 8\lambda'/\Delta, \quad (31)$$

where Δ is the separation of the A_2 and T_2 levels in the d^3 portion of fig. 2. From the g factors in table 3 and from values of Δ one may obtain values of λ' for the several ions. Alternatively, from the measured g factors in MgO, CaO and SrO for V^{2+} and Cr^{3+}, one may find the ratios of the quotients λ'/Δ for CaO and for SrO relative to MgO. Because of the larger lattice constants (0·482 mm in CaO and 0·508 nm in SrO versus 0·420 in MgO) one expects smaller values of Δ in CaO and SrO and hence an increased value of λ'/Δ. The spin Hamiltonian for d^3 ions in octahedral symmetry is:

$$\mathcal{H} = \beta g \vec{B} \cdot \vec{S} + A\vec{I} \cdot \vec{S}. \quad (32)$$

The spectrum of Ti^{1+} in CaO shows excellently resolved hyperfine splitting both from ^{47}Ti and ^{49}Ti. The former has $I = \frac{5}{2}$ and the latter $I = \frac{7}{2}$; however, the ratios μ/I (of nuclear magnetic moment to spin) are very nearly identical. Hence ordinarily one sees a hyperfine octet of eight lines, of which the central six come both from ^{47}Ti and ^{49}Ti. In this case, the lines are narrow enough so that one can see a small splitting on the central six lines. Naturally, there is an intense central line from the approximately 87% of Ti with zero nuclear spin. Here $^{47}A = {}^{49}A = 10\cdot 8 \times 10^{-4}$ cm^{-1}. ^{51}V constitutes 99·76% of all vanadium, and a hyperfine octet of unequal spacing is observed for ^{51}V^{2+} because the large hyperfine coupling introduces significant second-order corrections. Table 3 indicates that the coupling constant in MgO is less than in CaO. The natural abundance of ^{53}Cr is 9·54% and most of the octahedral ESR line intensity is concentrated in the central line from nuclides 50, 52 and 54. As chromium is also found in tetragonal or in rhombic symmetry (table 3, and § 3.4, § 3.5), the hyperfine quartet is a valuable means of identifying weak Cr^{3+} lines. The increasing hyperfine coupling from MgO through to SrO presumably reflects a reduction in covalent bonding. The ^{55}Mn^{4+} sextet spectrum has the highest g factor of any of the d^3 ions in MgO. Again the hyperfine coupling in calcium oxide is higher than in magnesium oxide (table 3).

ESR in the excited ^{2}E state of the 3d^3 ions V^{2+}, Cr^{3+} (Chase 1968) and Mn^{4+} (Henderson, to be published) has been observed. A steady state population of ^{2}E is achieved by continuous excitation in the broad ^{4}T$_2$ and ^{4}T$_1$ absorption bands, from which excited ions decay non-radiatively into the ^{2}E level. This is followed by radiative decay to the ground state via the R-line (^{2}E → ^{4}A$_2$) fluorescence (see § 3.4). In a magnetic field there is a splitting of both ground and excited states due to the electronic Zeeman interaction, and as a consequence there are six emissive transitions. However, each of the excited-state levels, $|\pm\tfrac{1}{2}\rangle$ emits preferentially one sense of polarization. Consequently at low temperature (1·6°K) sampling the luminescence parallel to the magnetic field with a circular polarization analyser yields a nett circular polarization on account of the Boltzmann-like distribution of the $|\pm\tfrac{1}{2}\rangle$ spin components of ^{2}E. Since the populations $n(+\tfrac{1}{2})$ and $n(-\tfrac{1}{2})$ change at resonance, ESR can be detected by monitoring the resultant change in the circular polarization. In each of the cases, V^{2+}, Cr^{3+} and Mn^{4+}, the optically detected resonance shows a sample dependent anisotropic broadening due to the influence of local random strain on the orbitally degenerate ^{2}E level. Large g-shifts are observed due to configuration mixing and spin-orbit coupling in this state, which are in agreement with the hfs observed for V^{2+} and Mn^{4+}.

Cr^3 and Mn^4 ions are observed in tetragonal symmetry sites in MgO and CaO: the distortion is due to a cation vacancy in the nearest cation site along a $\{100\}$ direction (fig. 12). The lowering of symmetry consequent upon the presence of the cation vacancy removes partially the electronic spin degeneracy as discussed in § 2.1, separating the two doublets $M_s = \pm\tfrac{1}{2}$ and $\pm\tfrac{3}{2}$ by an amount $2D$. If $2D$ is small compared with the electronic Zeeman energy then the resonance fields are easily calculated to second order in perturbation theory. For $\vec{B}$ parallel to the tetragonal axis the transitions $M_s \leftrightarrow M_s - 1$ are observed at fields

$$B = \frac{h\nu}{g\beta} - 2D(M_s - \tfrac{1}{2}) \tag{33}$$

and for $\vec{B}$ perpendicular to the axis

$$B = \frac{h\nu}{g\beta} + D(M_s - \tfrac{1}{2}) - \frac{D^2}{16B}(9 - 12M_s(M_s - 1)) \tag{34}$$

Table 3. Spin Hamiltonian parameters of 3d^3 ions in alkaline earth oxides†

Ion	Host	Symmetry	Temp °K	g	A	D	E	Reference
Ti$^+$	MgO	cubic	77	1·9925	$^{47,49}A = 10·8$	—	—	Tomlinson and Henderson (1968)
V^{2+}	MgO	cubic	290	1·9800	$^{51}A = 74·3$			Kolopus et al. (1965)
	CaO	cubic	77	1·9683	76·15			
	SrO	cubic	77	1·9592	79·7			
Cr^{3+}	MgO	cubic	290	1·9800	$^{53}A = 16·0$			
	CaO	cubic	77	1·9732	17·0			
	SrO	cubic	77	1·9663				
			4·2	1·9686	17·3			
Mn^{4+}	MgO	cubic	290	1·9942	$^{55}A = 70·8$			Henderson and Hall (1967)
	CaO	cubic	77	1·9924				
Cr^{3+}	MgO	tetragonal	77	$g_\parallel = 1·97854$	$^{53}A = 16·2$	819·2		
				$g_\perp = 1·98171$				
	CaO	tetragonal	77	$g_\parallel = 1·9697$		1360·6		
				$g_\perp = 1·9751$				
Mn^{4+}	MgO	tetragonal	90	$g_\parallel = 1·9931$	$^{55}A = -71·1$	$-528·7$		Davies et al. (1969)
				$g_\perp = 1·9940$	$A_\perp = -70·6$			
V^{2+}	MgO	rhombic*	290	1·9910	$^{53}A = 72·0$	±0·0281	±0·1119	Codling and Henderson (1971)
						0·0580	±0·1126	Davies and Wertz (1974)
Cr^{3+}	MgO	rhombic*	290	1·9800		±0·0339	±0·2511	Wertz and Auzins (1957, 1967)
						±0·031	±0·22	Griffiths and Orton (1959)
						±0·026	±0·226	Codling and Henderson (1971)

† Values of A, D and E are to be multiplied by 10^{-4} cm^{-1}.
* The axis of the system are taken to be parallel to [001], Z-axis; [110], X-axis and [$\bar{1}$10], Y-axis.

g being assumed isotropic. In the case of Cr^{3+} a relatively simple spectrum is observed (Wertz and Auzins 1957), which for $\vec{B}$ parallel to [100] comprises the six line resonance illustrated in fig. 13. As in the case of the axial Ti^{3+} spectrum discussed above, the doubly intense lines correspond to centres with a tetragonal axis perpendicular to the applied field. A similar spectrum is observed for Cr^{3+} ions in CaO. An extensive hyperfine structure is to be seen for Mn^{4+} in MgO due to the 100% abundant isotope ^{55}Mn. Including terms in D in the energy denominators of the various terms of the eigenvalues makes it possible to show that in this case D and A are of the same sign (Davies et al. 1969). Depopulation experiments at $1.6°K$ show D, and therefore A, to be negative.

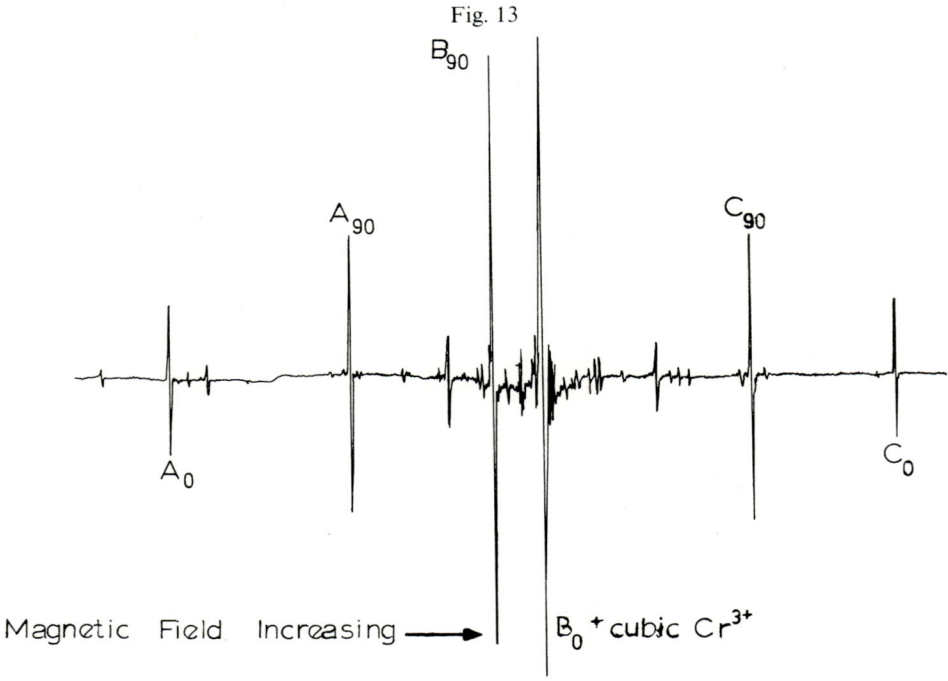

Fig. 13

ESR spectrum of Cr^{3+} in tetragonal symmetry in MgO at X-band frequency. The lines labelled A_0, B_0, C_0 are due to ions with tetragonal axis parallel to the applied field B_0, whereas A_{90}, B_{90} and C_{90} are doubly intense lines due to ions with symmetry axis perpendicular to a [100]-direction. (After Wertz and Auzins 1957.)

In many commercially available crystals of MgO and CaO the tetragonally symmetric Cr^{3+} spectrum is accompanied by weaker satellite lines. The relative intensities of all the tetragonal spectra are strongly sensitive to thermal treatment, and particularly sensitive to the presence of Al^{3+} and Ga^{3+} in the crystals, (Henderson and Hall 1967, Wertz and Auzins 1967, O'Donnell unpublished). Similar satellite lines are observed for the orthorhombic Cr^{3+} spectrum in MgO. In general, the g-values remain unchanged for the satellite spectrum. However, the separation between lines is less than in the normal tetragonal spectrum, suggesting that the satellite spectra arise from Cr^{3+} ions with a smaller axial component of the crystal field: the values of D are reduced from those given in table 3 to -798.3×10^{-4} cm^{-1} and in MgO and -1268.8×10^{-4} cm^{-1} in CaO. The spectra have been attributed to a charge neutral configuration in which the nearest and next-nearest neighbour cation sites along a [100] axis are respectively vacant and occupied by a diamagnetic tripositive ion, the most likely ions being Al^{3+} and Ga^{3+}.

In addition to the EPR spectrum due to Cr^{3+} ions in tetragonal symmetry, in MgO one usually observes a distinct spectrum arising from Cr^{3+} ions with axis of symmetry parallel to [110] crystal directions. This symmetry is ascribed to the effects of a cation vacancy in a position nearest the central paramagnetic ion. The spectrum was first reported by Wertz and Auzins (1957) and analysed in detail using data taken at X- and K-band frequencies by Griffiths and Orton (1959). Codling and Henderson (1971) who used Q-band frequencies to detect all possible transitions in the spectrum, obtained good agreement with the results of the earlier workers. In addition, they observed a spectrum due to V^{2+} ions in orthorhombic site, which they ascribed to the analogous centre in V-doped MgO. Recently Davies and Wertz (1975) have observed two further spectra due to V^{2+} in orthorhombic sites in crystals of undoped MgO subsequent to X-irradiation. The centres are unstable at room temperature and are presumed to be formed by ionization of V^{3+} ions associated with cation vacancies. In the case of the weaker spectrum it is suggested that a second impurity ion is associated with the complex. Experimentally using ESR one cannot distinguish between the models responsible for the three different orthorhombic spectra.

In MgO crystals containing a large concentration of V^{2+} ions an EPR spectrum is observed due to pairs of V^{2+} ions in next nearest neighbour sites (Codling and Henderson 1971b). The spectrum has symmetry about a [100] crystal axis. Due to magnetic exchange interaction $J\vec{S}_1 . \vec{S}_2$ between the two ions with effective electronic spins $S_1 = S_2$ there is a coupling such that the total spin $\vec{S} = \vec{S}_1 + \vec{S}_2$ can take values $S_1 + S_2, S_1 + S_2 - 1, \ldots 0$. For a $3d^3$ ion there are four total spin states $S = 3, 2, 1, 0$, the splittings between which are given by a Landé interval rule. The ESR spectrum is interpreted by considering transitions within each total spin state separately. The hyperfine structure must also be evaluated within the coupled representation: the interaction takes the form $A/2S . (I_1 + I_2)$. This gives a characteristic 15 line hyperfine structure, which since $I_1 = I_2 = \frac{7}{2}$ has an intensity distribution $1:2:3:\ldots 8 \ldots 3:2:1$ corresponding to the ways of compounding each M_I value. The splitting between lines is $A/2 = 37 \times 10^{-4}$ cm^{-1}, half the value found for isolated V^{2+} ions in MgO. The value of the exchange constant J is found from the temperature dependence of the $S = 2$ state to be ~ 60 cm^{-1}, although a small biquadratic term $j(S_1 . S_2)^2$ is needed to accurately describe the overall behaviour. Furthermore the measured dipole–dipole interactions are substantially smaller than is calculated even when covalent electron spin transfer between magnetic and ligand ions is assumed. A pair spectrum due to Cr^{3+} ions in MgO has been reported (Marguglio and Kim 1975). The spectrum is extremely complex and its analysis almost certainly incorrect (Henderson and O'Donnell to be published).

For the d^5 ions one must be concerned with the parameter $3a$, which measures the zero-field splitting of the $\pm\frac{1}{2}$ levels from the $\pm\frac{3}{2}$ and $\pm\frac{5}{2}$ levels which are degenerate. The appropriate spin Hamiltonian is;

$$\mathcal{H} = g\beta \vec{B} . \vec{S} + A\vec{S} . \vec{I} + \frac{a}{6}[S_x^4 + S_y^4 + S_z^4 - \tfrac{1}{5}S(S+1)(3S^2 + 3S + 1)]. \quad (35)$$

For Mn^{2+} in MgO, which has the highest value of a ($19\cdot01 \times 10^{-4}$ cm^{-1}), the zero-field splitting is several-fold smaller than the hyperfine interaction, and hence one

Table 4. Spin Hamiltonian parameters of $3d^5$ ions in alkaline earth oxides

Ion	Host	Symmetry	Temp °K	g	A $\times 10^{-4}$ cm^{-1}	D $\times 10^{-4}$ cm^{-1}	a $\times 10^{-4}$ cm^{-1}	Reference
Mn^{2+}	MgO	cubic	290	2·0010	−81·11	—	91·61	
	CaO	cubic	290	2·0011	80·7			
		cubic	77		81·6			
		cubic	20		−81·7		6·00	
	SrO	cubic	290	2·0012	−78·7		4·30	
			77	2·0014	80·2			
Fe^{3+}	MgO	cubic	290	2·0033	10·1		203·8	
	CaO	cubic	77	2·0059			−64·3	
			22	2·0059	10·5		−65·1	
Mn^{2+}	BaO	trigonal	77	$g_\parallel = 2\cdot0017$ $g_\perp = 2\cdot0013$	$A_\parallel = 67\cdot5$ $A_\perp = 68\cdot5$	778·0		Weightman et al. (1971)
Fe^{3+}	MgO	tetragonal	290	$g_\parallel = 2\cdot0026$ $g_\perp = 2\cdot0037$		4700	26·2	Henderson et al. (1971 a)

sees a group of six pentads. The central lines (i.e. the $M_S = +\frac{1}{2} \leftrightarrow -\frac{1}{2}$ transitions) are essentially isotropic. There is a small increase in the ^{55}Mn hyperfine coupling on lowering the temperature (table 4), and there are relatively small changes in hyperfine couplings in the various oxides. For Fe^{3+}, the value of a is roughly ten times as large as for Mn^{2+}, and the outer pairs of the pentad show large anisotropy. The $M_S = +\frac{1}{2} \leftrightarrow -\frac{1}{2}$ transition shows an anisotropy of the order of 15 G at X-band in MgO. The value of a for Fe^{3+} in calcium oxide is less than one-third that in magnesium oxide. The ratio of a values for Mn^{2+} in the two hosts roughly parallel that for Fe^{3+} in the two crystals. These results imply a 30% smaller crystalline electric field in calcium oxide than in magnesium oxide (Low and Rubins 1963). Although the natural abundance of ^{57}Fe is only 2·245%, the hyperfine doublet has been seen about the $M_S = +\frac{1}{2} \leftrightarrow -\frac{1}{2}$ transition under favourable conditions in both magnesium and calcium oxide (table 4).

The ferric ion has also been observed in a distorted site in MgO, the distortion apparently arising from the next nearest neighbour site being vacant. Studies at X- and Q-band reveal that the transitions, both allowed and forbidden, are fitted by the spin Hamiltonian

$$\mathcal{H} = \mu_\beta \vec{B} \cdot \tilde{g} \cdot \vec{S} + D\{S_z^2 - \tfrac{1}{3}S(S+1)\} + \frac{a}{6}\{S_x^4 + S_y^4 + S_z^4 - \tfrac{1}{5}S(S+1)(3S^2 + 3S + 1)\}$$

with $g_\parallel = 2·0026$, $g_\perp = 2·0037$, $a = 0·00262$ cm^{-1} and $D = +0·470$ cm^{-1} (Henderson et al. 1971 a).

In BaO a non-cubic Mn^{2+} has been observed (Weightman et al. 1971): the spectrum displays trigonal symmetry about the {111} axis. The distortion may be due to relaxation of the Mn^{2+} ion into off-centre positions along {111} directions. Very detailed studies of Mn^{2+} pairs in both MgO and CaO have been reported (Coles et al. 1960, Harris and Owen 1963, Harris 1973). Large deviations from the Landé interval rule are observed which are due to exchange striction: crystal field and magnetic dipole–dipole interactions are also modified by the magnetostrictive interaction (Harris 1973). The exchange interaction J, anisotropic interaction D_E and crystal field parameter D_C all vary in a way consistent with the varying ionic separation. The results give good agreement with the magnetic properties deduced for MnO.

For a $3d^6$ ion (e.g. Fe^{2+}), the 5 × 3-fold spin-plus-orbital degeneracy of the T_2 level of fig. 4 is partially lifted, such that the lowest level is a triplet. In succession above this level are a doublet, triplet, singlet and two triplets. The proximity of these upper levels leads to a short value of the spin-lattice relaxation time T_1, hence detection of ESR absorption requires temperatures of the order of 20°K or lower. An effective spin $S' = 1$ appropriate to the lowest level is used in the simple spin Hamiltonian which involves only the isotropic Zeeman interaction. Here the value of g computed from the spin–orbit interaction is given by (Low and Weger 1960):

$$g = 3(1·0011) + \frac{k}{2} + \frac{9}{25}\frac{\lambda'}{\Delta}.$$

At 20°K, one observes an absorption band for Fe^{2+} at 10.850 cm^{-1} (Jones 1967). If this value is taken as Δ and $\lambda' = -100$ cm^{-1}, the value of the orbital reduction factor k computed from the experimental g value of 3·4277 (table 5) is about 0·78. The great width of the Fe^{2+} line is indicative of distortions from octahedral symmetry, as is also the $\Delta M = 2$ line referred to in § 3.1.1.

Table 5. Spin Hamiltonian parameters of d^6–d^9 ions in alkaline earth oxides

Ion	Host	Symmetry	Temp °K	g	A 10^{-4} cm^{-1}	Reference and comments
Fe^{2+}	MgO	cubic	4·2	3·4277		Double quantum line on broad line
	CaO	cubic	2	3·298		Double quantum line on broad line
Fe$^+$	MgO	cubic	4·2	4·1307		
	CaO	cubic	4·2	4·1579		
Co^{2+}	MgO	cubic	20	4·2785	3·39	
	CaO	cubic	20	4·3747	96·85	
Ni^{3+}	MgO	cubic	4·2	$g_{iso} = 2·1685$	131·5	Dynamic Jahn–Teller effect (Schoenberg et al. 1972)
	CaO	cubic	77	2·2814		Originally identified as Ni$^+$, corrected by Höchli et al. (1965)
			<65	$g_\parallel = 2·0672$		
				$g_\perp = 2·3828$		
Co$^+$	MgO	cubic	77	2·1728	54·0	
Ni^{2+}	CaO	cubic	12	2·2756	31·6	
	MgO	cubic	77	2·2145	8·3	Double quantum line (Smith et al. 1969)
			4·2	2·234		
	CaO	cubic	20	2·327		
Cu^{2+}	MgO	cubic	1·2	$g_\parallel = 2·384$	$A_\parallel = 58·3$	Complex behaviour due to tunnel effects at low temperature (Coffman 1966, 1968, Coffman et al. 1968)
				$g_\perp = 2·096$		
	CaO	cubic	4·2	2·2223	29·1	Static J–T effect (Boatner et al. 1973)
Ag^{2+}	MgO	tetragonal	1·3	$g_\parallel = 2·112$	$^{107}A_\parallel = 28·79$	Intermediate J–T effect (Boatner et al. 1973)
				$g_\perp = 2·017$	$^{107}A_\perp = 22·92$	
	CaO	tetragonal	1·3	$g_\parallel = 2·166$	$^{107}A_\parallel = 2·94$	Dynamic J–T effect (Boatner et al. 1973)
				$g_\perp = 2·031$	$^{107}A_\perp = 21·0$	
	SrO			$g_1 = 2·0998$	$A_1 = \pm 25·8$	
				$qg_2 = +0·0563$	$qA_2 = \pm 5·4$	

For the d^7 ions (Fe^{1+}, Co^{2+}, Ni^{3+}) in MgO and CaO, the spin–orbit interaction splits the 4×3-fold spin-plus-orbital degeneracy such that the lowest level is a doublet. A quartet and a sextet lie respectively 600 and 1000 cm^{-1} above this doublet. These low-lying levels lead to a small value of T_1 because of spin–orbit coupling to the doublet. Hence temperatures near 20°K are required for observation of ESR spectra. The effective spin $S' = \frac{1}{2}$ for the lowest level, and hence the spin Hamiltonian for Co^{2+} is that given by eqn. (32). No hyperfine interaction has been detected for $^{57}Fe^{1+}$ and $^{61}Ni^{3+}$ in MgO. The eigenfunctions of the spin-degenerate ground states are (Low 1958):

$$|M_z = \pm\tfrac{1}{2}\rangle = (\tfrac{1}{2})^{1/2}|\mp 1, \pm\tfrac{3}{2}\rangle - (\tfrac{1}{3})^{1/2}|0, \pm\tfrac{1}{2}\rangle + (\tfrac{1}{6})^{1/2}|\pm 1, \pm\tfrac{1}{2}\rangle.$$

The isotropic g factor is obtained by the evaluation of:

$$g = 2\langle\pm\tfrac{1}{2}|L_z + 2S_z|\pm\tfrac{1}{2}\rangle = 4.33.$$

This first-order result is modified in second order by the addition of $-15\lambda'/2\Delta$, where Δ is the separation of the T_1 and T_2 levels. In magnesium oxide Δ is 9600 cm^{-1} for Co^{2+}, while for the free ion λ is -178 cm^{-1}. The experimental value of $g = 4.2785$ is low even without the second-order correction. This is attributed to an appreciable covalency, which requires an admixture of oxygen ion wave functions into the 3d transition ion orbitals. This reduces both the orbital contribution and the second-order effect of the spin–orbit coupling upon g. The orbital-reduction factor of Co^{2+} is given as 0.89 (Low 1958). A concomitant effect of fractional covalent bonding is the reduction of the hyperfine splitting. In MgO, $A = 96.8 \times 10^{-4}$ cm^{-1}, while in CaO, in which the lattice constant is 0.482 instead of 0.420 nm, the coupling constant is 131.5×10^{-4} cm^{-1} (Low and Rubins 1963). The value of $g(4.3737)$ for Co^{2+} in CaO also indicates a lower fractional covalent bonding than in MgO. The metastable ions Fe^{1+} and Ni^{3+} may be generated by X-irradiation in both MgO and in CaO (table 5). Too little is known of the properties of Fe^{1+} to interpret the significantly lower g factor than for Co^{2+}.

The Ni^{3+} ion in both MgO and in CaO had initially been identified as Ni^{1+}; however, Höchli et al. (1965) showed that the g factor of the single line produced by Li doping of MgO or Na doping of CaO is essentially identical with the single crystal value after X-irradiation. For CaO below 65°K, three sets of lines of tetragonal symmetry were obtained, with $g_\perp > g_\parallel$. This is taken as evidence of a static Jahn-Teller effect, with an elongation of the oxygen octahedron (Öpik and Pryce 1957). The Jahn–Teller splittings of the ground states of MgO and of CaO are given as 1300 ± 100 cm^{-1} and 2500 ± 200 cm^{-1} respectively (Höchli et al. 1965). The latter is said to be in a $t^6 e$ configuration. The much-reduced g factors (as compared with Co^{2+}) are in the range of the g components of Ni^{3+} in TiO_2 (Gerritsen and Sabisky 1962).

A recent study by Schoenberg et al. (1974) of Ni^{3+} in MgO crystals grown from a lead fluoride flux show an anisotropic doublet, in addition to the isotropic line reported in the earlier work. Analysis of the spectra shows that the anisotropic lines originate from a nearly isolated ground vibronic doublet split by large random strains, while the isotropic lines result from fast averaging of the anisotropic spectra.

The Ni^{2+} ion is the most important of the d^8 ions in magnesium oxide and calcium oxide, though the ion Co^{1+} may be produced by X-irradiation. Though the A_2 level of d^8 lies lowest in fig. 4, the spin of 1 makes possible a zero-field splitting in many hosts. Since one does see a symmetrical, though wide line in both oxides, the zero-field splitting parameter D will not be explicitly incorporated into the spin Hamiltonian. Relaxation properties of the system do give rise to a dip in the centre of the absorption line. This topic and that of double-quantum transitions are discussed briefly in § 3.1.2. The g values and spin Hamiltonian for $^{61}Ni^{2+}$ and $^{59}Co^{1+}$ in MgO and CaO are given respectively by eqns. (1) and (2). For the free ion Ni^{2+}, $\lambda = -322 \text{ cm}^{-1}$, and in MgO, $\Delta = 8150 \text{ cm}^{-1}$ (Pappalardo et al. 1961). At 77°K, $g = 2.2145$, suggesting that $\lambda' = 0.7\lambda$. Again, the isoelectronic monovalent ion (Co^{1+}) has a significantly lower g factor ($g = 2.1728$) than the divalent ion, in parallel with the behaviour of Fe^{1+}–Co^{2+} pair.

The d^9 ion Cu^{2+} in a presumably octahedral field has a four-fold spin-plus-orbital degeneracy which is unaltered by spin–orbit coupling. However, octahedral Cu^{2+} is unstable and undergoes a relatively large Jahn–Teller distortion, which is usually an elongation of the octahedron. The splitting is large enough that one readily sees ESR absorption at 77°K; again the appropriate spin Hamiltonian is that in eqn. (32). At this temperature in MgO, the factor $g = 2.192$ and the hyperfine coupling $A = 19.0 \times 10^{-4} \text{ cm}^{-1}$ are isotropic (Orton et al. 1961, Coffman 1968). The isotropic value of g is given by $g = 2 + 4u$, where $u \approx \lambda/\Delta$ (Abraham and Pryce 1950, O'Brien 1964); thus $u = 0.0480$ at 77°K. The isotropic behaviour is presumably the result of vibrational excitation which is sufficient to cause rapid reorientation among the equivalent sets of axes. However, in the region below 4.2°K, the behaviour is complicated by the appearance of a number of anisotropic spectra. At 1.2°K, the anisotropy of the observed lines is in poor agreement with a model of static tetragonal distortions which would leave a Kramers doublet lowest; for this, one expects $g_\parallel = 2 + 8u$ and $g_\perp = 2 + 2u$ (Coffman 1968). However, if the height of the potential barrier between the three equivalent tetragonal distortions is such that tunnelling ('inversion') is probable, the interaction between these states will lead to a Kramers doublet and a Kramers quartet, of which the latter lies lower (Bersuker 1963, 1965, O'Brien 1964). The behaviour of the observed lines at 1.2°K in MgO is consistent with the predictions based on the tunnelling model (Coffman 1968). Parallel experiments on CaO at 1.2°K again fit the tunnelling model, with a splitting of 0.006 cm^{-1} between the Kramers quartet and doublet (Coffman et al. 1968). The results on CaO confirm that the tetragonal distortions are compressions, as indicated by earlier work on line widths (Low and Suss 1963).

Other d^9 ions to show Jahn–Teller effects are Ni^{1+} (Schoenberg et al. 1974) and Ag^{2+} (Boatner et al. 1973). The spectrum from Ni^+ ions consists of an isotropic line and an anisotropic doublet similar to the Ni^{3+} spectrum, due to a strain split vibronic doublet with a small admixture of an excited singlet. EPR studies of Ag^{2+} in the series MgO and CaO and SrO shows a systematic transition from a predominantly dynamic Jahn–Teller effect (MgO) to a static effect in SrO (Reynolds et al. 1974, Boatner et al. 1973, Reynolds and Boatner 1976).

Improved crystal growth techniques has made it possible to incorporate a number of rare earth ions in the alkaline earth ions (Butler et al. 1971, Abraham et al. 1971). The ESR spectrum of Eu^{2+} and Gd^{3+} has been observed in MgO, CaO, SrO and BaO for the ions in cubic symmetry sites. Gd^{3+} has been observed in orthorhombic symmetry sites in CaO (McGeehin and Henderson 1974). Dy^{3+}, Er^{3+} and Yb^{3+} has been successfully incorporated in MgO, CaO and SrO and ESR spectra reported. The data are collected in table 6. Tm^{2+} may be observed in Tm:CaO following X-irradiation below 77 K.

3.1.1. *Forbidden lines in impurity spectra*

The most frequently observed forbidden transitions of impurity ions in the alkaline earth oxides are those for which $\Delta M_S = \pm 1$ and $\Delta M_I = \pm 1$ (for an ion with $I \neq 0$). The best known examples are a pair of lines which occur between each pair of the Mn^{2+} sextet lines corresponding to $M_S = \frac{1}{2} \leftrightarrow -\frac{1}{2}$. They have most frequently been observed in hosts which contribute an axial field component (see references in Smith *et al.* 1968); they are also seen in non-oriented solids such as modelling clay. These lines arise because an $|M_S, M_I\rangle$ state has admixed into it some of the states $|M_S, M_I \pm 1\rangle$ as a result of the combined effects of the octahedral field and the hyperfine operators, viz. $a/6[S_x^4 + S_y^4 - S_z^4]$ and $A\mathbf{I} \cdot \mathbf{S}$. In an octahedral crystal the intensity of these lines is proportional to $(\sin 4\theta)^2$, where θ is the angle between the field and a principal axis of the crystal (Drumheller and Rubins 1964). The lines have also been studied in CaO (Drumheller 1964, Tanimoto and Kemp 1966) and in SrO (Tanimoto and Kemp 1966). Some additional lines are seen if the microwave and the static fields are parallel (Smith *et al.* 1968).

In the ESR spectrum of Fe^{2+} there is a line which occurs at approximately one-half the normal resonant field, i.e. at $g = 6\cdot83$ (Low 1958). The $\Delta M_S = 2$ transition which gives rise to this line, is strictly forbidden in an exactly octahedral field. Random strains from dislocations, from other line imperfections, or from charge abnormalities lead to mixing of states, and hence the transition between them becomes allowed. The derivative line shape resembles more nearly that of a skewed absorption line, having a steep slope on the high-field side. The corresponding $\Delta M_S = 2$ line in CaO is also unsymmetrical (Shuskus 1964, Low and Weger 1963). It is found that at X-band the shape of the line can be reproduced by assuming a random distribution of distortions and hence of zero-field splittings. For a wave length of 4 mm, the line shape is explained in terms of a strain-induced shift in the g factor (McMahon 1964).

In addition to $\Delta M_S = 2$ transitions, Kolopus and Holroyd (1965) have observed transitions corresponding to $\Delta M_S = 3$, $\Delta M_S = 4$ and $\Delta M_S = 5$ in the low-field ESR spectrum of Fe^{3+} in magnesium oxide. The angular dependence of these lines fits excellently the computed curves when the octahedral field splitting parameter $a = 209\cdot0 \times 10^{-4}$ cm^{-1}. This value is somewhat higher than the value $203\cdot8 \times 10^{-4}$ cm^{-1} given in table 4.

3.1.2. *Unusual aspects of impurity spectra in MgO*

Certain extra lines which are not accounted for as single-quantum transitions and which are generally narrower than the allowed lines, make a striking appearance in some of the iron-group substitutional-ion spectra in magnesium and calcium oxides. Perhaps the first such lines observed were the double-quantum transitions between the $-\frac{3}{2} \leftrightarrow +\frac{1}{2}$ and the $-\frac{1}{2} \leftrightarrow +\frac{3}{2}$ levels; these are seen as a pair of lines symmetrically spaced about the centre of each pentad, both at room temperature and at 77°K. At the latter temperature, the triple-quantum transitions between the $-\frac{1}{2} \leftrightarrow +\frac{5}{2}$ and the $-\frac{5}{2} \leftrightarrow +\frac{1}{2}$ levels were also observed. Second-order time-dependent perturbation theory was used to obtain the ratio of the two-quantum to the single-quantum transitions. Apparently the intensity of the double-quantum line in this simple case should be half that of the single-quantum line (Sorokin *et al.* 1958).

Rather more striking is the very sharp double-quantum line in the centre of the normal Fe^{2+} ESR line, which is at least a 100-fold broader (Low 1958). The origin

Table 6. ESR parameters of Rare-Earth Ions in Cubic Symmetry*

Ion	Host	Temp °K	g	A $\times 10^{-4}$ cm^{-1}	c $\times 10^{-4}$ cm^{-1}	d $\times 10^{-4}$ cm^{-1}	References
Eu^{2+}	MgO	77	1·9894	$^{151}A = -31·6$ $^{153}A = -13·95$	−291·8	+14·7	Abraham et al. 1971
	CaO	4·2	1·9894	$^{151}A = 30·09$	−292·6	+14·6	
		77	1·9917	$^{153}A = 13·42$	−100·4	−8·4	
	SrO	1·6–77	1·991	$^{151}A = 30·1$ $^{153}A = 13·4$	2·0	2·0	
	BaO	4·2	1·9915	$^{151}A = 29·6$ $^{153}A = 13·1$	+76		
Gd^{3+}	MgO	77	1·9920	$^{155}A = 3·85$ $^{157}A = 5·06$	−139·8	+8·8	Tomlinson and Henderson (1968)
	CaO	4·2	1·9920		−140·0	+8·8	Abraham et al. (1967)
		77	1·9908	$^{155}A = 3·60$ $^{157}A = 4·67$	+48·4	−4·64	Kolopus et al. (1966)
	SrO	4·2	1·993		+49·1	−4·70	
		77·4	1·9918		21·6	3·3	
	BaO	4·2	1·989		23·2		Mann and Holroyd (1968)
Dy^{3+}	MgO	4·2	6·539	$^{161}A = 184$ $^{163}A = 257$			Reynolds et al. (1974)
	CaO	4·2	6·566	$^{161}A = 182$ $^{163}A = 260$			
	SrO	4·2	6·572	$^{161}A = 184$ $^{163}A = 261$			
	BaO	4·2	6·633	—			

Table 6—(continued)

Ion	Host	Temp °K	g	$A \times 10^{-4}$ cm^{-1}	$c \times 10^{-4}$ cm^{-1}	$d \times 10^{-4}$ cm^{-1}	References
Er^{3+}	MgO	20	[100] 4·62 [110] 3·86 [111] 3·60				
	CaO	20	[100] 4·84 [110] 3·85 [111] 3·50				
Yb^{3+}	MgO	4·2	2·5662	$^{171}A = 680·7$ $^{173}A = 187·5$			Reynolds et al. (1974)
	CaO	4·2	2·586	$^{171}A = 688$ $^{173}A = \quad A$			
	SrO	4·2	2·591	$^{171}A = 685$ $^{173}A = 190$			
	BaO		2·5926	$^{171}A = 692·4$ $^{173}A = 192$			
Ce^{3+}	CaO	4·2	0·7963				Reynolds et al. (1972)
	SrO	4·2	0·8948				
	BaO	4·2	0·9340				

* Lower symmetry sites have been identified in the case of Gd^{3+} : CaO, (McGeehin et al. 1974), Yb^{3+} IN CaO (Reynolds et al. 1974).

of the narrow line was suggested by the parallel behaviour of the double-quantum line for Ni^{2+} in MgO (Orton *et al.* 1960 b). For this ion the effective spin is also 1. The most obvious test for a double-quantum line is its microwave-power dependence. It should not be seen at all at low power; for high power, one may readily predict that the intensity will be proportional to the square of the power. However, in practice one must take into account the particular design of the spectrometer employed. Specifically, if a crystal detector is employed, then the biasing power should be kept constant. It is readily possible to achieve this condition and still allow the incident power on the sample to vary greatly. This is accomplished by the use of one attenuator ahead of and one just beyond the circulator connecting the sample cavity. In this system, as one decreases the attenuation ahead of the cavity, one increases it behond the cavity to keep the bias constant. Under these conditions of operation, the ESR line intensity for a normal $\Delta M_S = 1$ transition increases as the square root of the power incident on the cavity. For a double-quantum transition the intensity of the line increases linearly with the applied power in this system. The intensity of the double-quantum line of Ni^{2+} in MgO does indeed follow this dependence. There is a second very different and definitive test which makes use of a bimodal cavity, e.g. a TE_{111} rectangular cavity having two resonant frequencies within tens of Mhz of one another. The crystal is placed in a region of the cavity which has a large microwave magnetic field from both modes. The cavity is attached to two independent detection systems, and the signals are recorded from each channel, but a pair of these will occur at the same field value. These correspond to the absorption of one quantum of each of the two frequencies. The signal at a field value corresponding to the lower frequency shows the absorption of two quanta of that frequency. Symmetrically spaced to higher field is the line from the transition in which two of the quanta of higher frequency were absorbed (Orton *et al.* 1960 b). Double-quantum transitions are observed in calcium oxide for both Ni^{2+} (Low and Rubins 1963) and Fe^{2+} (Shuskus 1964). Subsequently microwave acoustic resonance has also been used to verify the double-quantum nature of the central sharp line of Fe^{2+} (Shiren 1961).

The existence of a double-quantum transition made it readily possible to identify the Co^{1+} ion produced by X-irradiation. The normal $\Delta M_S = 1$ lines are very broad for Co^{1+} for the same reason that the Ni^{2+} and Fe^{3+} lines are broad; that is, there are strains which give rise to a distribution of zero-field splittings of both signs and of varying magnitude. This zero-field splitting may be regarded as causing a shift of the $|0\rangle$ state, while the separation of the states which at high field may be designated as $|+1\rangle$ and $|-1\rangle$ are essentially unaffected. Since the abundance of ^{59}Co ($I = \frac{7}{2}$) is $\approx 100\%$, the spectrum of Co^{1+} at high microwave power consists of eight sharp double-quantum lines superimposed upon eight much broader lines (Orton *et al.* 1960 a).

These double-quantum lines are intriguing, but they can hardly be called astonishing. The latter adjective seems appropriate for two narrow lines in the spectra of Fe^{3+} and Mn^{2+}, even though their double-quantum nature has been unambiguously established (Auzins and Wertz 1967). For Fe^{2+}, Ni^{2+} and Co^{1+} (especially for a two-quantum process with different frequencies), one might imagine an almost simultaneous process of reaching a virtual level by the first quantum and then attaining the final level by absorption of the second quantum. However, for the $3d^5$ ions, *the pair of intermediate levels involved in the two-quantum process differ by 2 in their M_S values*! Moreover, the sets of levels involved do *not* have the same spacings, unlike the $3d^6$ or $3d^8$ ions, in the absence of strain-induced distortions. Separate uniaxial-stress experiments have shown that the $\pm\frac{5}{2} \leftrightarrow \pm\frac{3}{2}$ lines are shifted by twice as much as the $\pm\frac{3}{2} \leftrightarrow \pm\frac{1}{2}$ lines (Feher 1964). Since the more closely spaced pairs of levels undergo the greater shift, the separation of two pairs of levels will become equal at some value of the stress. No external stresses are applied here; however, internal strains result in a range of tetragonal distortions which produce a range of zero-field splitting parameters D. Thus distortions at some Fe^{3+} ions are of just the right magnitude to yield equality of spacing of two pairs of levels. It is required that there be no change in the parameter a. The constancy of a is supported by the unaxial stress experiments (Feher 1964). One predicts that one double-quantum line should be located at one-third the separation of the $-\frac{3}{2} \leftrightarrow -\frac{1}{2}$ and the $+\frac{3}{2} \leftrightarrow +\frac{5}{2}$ lines; the other should be located at one-third the distance between the $+\frac{1}{2} \leftrightarrow +\frac{3}{2}$ and the $-\frac{5}{2} - \frac{3}{2}$ lines. They are indeed found at these positions.

The double-quantum nature of the lines is further verified by the criteria outlined earlier. First, in the two-attenuator spectrometer system described earlier, the signal intensity for the two narrow lines increases in proportion to the power, as it should for a double-quantum transition. Secondly, the bimodal cavity system was again used. For the $3d^5$ system, one expects three double-quantum lines from each channel and a total of four distinct lines. There are now two intercombination lines which are recorded by both channels, for now the resonance-field for $h\nu_a + h\nu_b$ depends upon whether $h\nu_a$ corresponds to the transition between the lower or upper pairs of levels (e.g. $-\frac{3}{2} \leftrightarrow -\frac{1}{2}$ or $+\frac{3}{2} \leftrightarrow +\frac{5}{2}$). In the two-frequency experiment, the line at lowest field corresponds to two quanta at ν_a, and that at highest field to two quanta at ν_b. The four lines are uniformly spaced, as expected.

For the high-field double-quantum transition the transitions occur between the $-\frac{5}{2} \leftrightarrow -\frac{3}{2}$ and the $+\frac{1}{2} \leftrightarrow +\frac{3}{2}$ levels. One may perform the same bimodal cavity experiment and get another set of four equally spaced lines (but with a different spacing of predicted magnitude). Thus the two sharp lines of the usual spectrum satisfy in detail the several tests for double-quantum transitions. The same phenomena are seen for Mn^{2+}, the spectrum of which is essentially a six-fold reduplication of that of Fe^{3+}. There appear to be two alternatives: (1) By some process not involving microwave radiation, a single ion is caused to effect a transition between levels differing by two units in M_S after absorbing the first quantum of microwave energy. (2) The transitions occur in nearby Fe^{3+} ions which both have the same value of D and are initially in the proper M_S states. Although the mechanism for (1) is not apparent, it would appear more promising than (2) (Auzins and Wertz 1967).

At low powers, such that the double-quantum line is not observed, the Ni^{2+} line is MgO has an unusual aspect. In the centre of the derivative presentation there appears to be a line of reversed phase (Orton *et al.* 1960 b). This apparent line at 135°K has a width approximately six-fold that of the double quantum line but less than one-tenth that of the 'normal' line. This 'inverse line' represents a dip in the centre of the integrated absorption line, and is not to be taken as a real line. The behaviour of Ni^{2+} (Low and Rubins 1963) and of Fe^{2+} in CaO is similar (Shuskus 1964). One may account quantitatively for the intensities of the normal, the inverse, and the double-quantum lines on the basis of a density-matrix treatment of relaxation processes in the system. Assuming a Lorentzian distribution of zero-field splittings ε, one expects symmetrical pairs of narrow $|\pm 1\rangle \leftrightarrow |0\rangle$ line components at $\pm\varepsilon$ if $\varepsilon \gg 1/T_2$. Hence $1/T_2$ is the transverse relaxation rate appropriate to the pair of spin packets with the same value of ε. If a pair of packets have $\varepsilon = 1/T_2$, they will lie near the centre, will partially overlap and will hence be broadened. This broadening has the effect of transferring intensity away from the centre of the line. The width of the apparent inverse line is then of the order of $1/T_2$. With increasing microwave power, the appearance of the double-quantum line will 'fill in' the centre of the line (Smith, Dravneiks and Wertz, 1968). It is believed that this behaviour is characteristic of any inhomogeneously broadened, symmetrical, three-level system.

3.2. *Effects of heat treatment*

It is presumed that an oxide crystal has initially been heated in a reducing atmosphere (e.g. H_2 at 1500°K) so as to provide a reference state. Under these conditions, the iron is present almost entirely as the Fe^{2+} ion. Most of the chromium is present as Cr^{3+}, since it is less affected by the reducing atmosphere. The most important effect of the heating for chromium is to determine the extent of association with vacancies. If cooling is slow, there will be a large fractional association, giving both singly and doubly associated centres (§ 3.4). Rapid quenching gives more isolated vacancies. The manganese is essentially all present in the form of Mn^{2+}, irrespective of the treatment, provided that monovalent ions have not been added. (It is not difficult to do this with single crystals grown from the melt; with modest doping levels ESR studies show that Mn^{4+} is stable.) The three impurities thus far mentioned are the predominant ones which may be detected by their ESR absorption in most crystals of MgO which have ever been prepared. If vanadium is present, a significant fraction of it will be reduced to the V^{2+} state. The same treatment also produces Ti^{3+}; one may then detect the ESR spectrum of those ions which are associated with a nnn cation vacancy. Presumably therefore, the precursor was Ti^{4+}. Cobalt, nickel and copper appear normally to be in the divalent state and are unaffected by reducing treatment.

Heating in oxygen at 1500°K is known to increase significantly the fraction of Fe^{3+} in the crystal. The latter is readily monitored by the intensity of either the intense 4·3 or the 5·7 eV optical absorption bands. Heating in vacuum or in hydrogen at 1500°K almost completely eliminates these bands if the iron content is not unduly high. It is known that heating in oxygen at this temperature results in an oxygen takeup proportional to the square root of the time of treatment (Weber 1951). The oxygen presumably is adsorbed at the surface, dissociates, and by a process of positive-ion-vacancy and hole migration acquires electrons to become surface O^{-2} ions. The electrons ultimately come from divalent ions (mostly Fe^{2+}), which become trivalent.

3.3. *Effects of ionizing radiation*

Given the reference state referred to in § 3.2, then e, γ or X-irradiation appears to induce the following changes:

$$Ti^{4+} \to Ti^{3+}$$
$$V^{3+} \to V^{2+}$$
$$Cr^{3+} \to Cr^{2+}$$
$$Fe^{3+} \leftarrow Fe^{2+} \to Fe^{1+}$$
$$Co^{2+} \to Co^{1+}$$
$$Ni^{3+} \leftarrow Ni^{2+}$$
$$Ru^{2+} \to Ru^{1+}$$
$$Pd^{3+} \leftarrow Pd^{2+}$$

A brief review indicates the degree of assurance with which these processes are acceptable. Observation of the ESR spectra of the final products Ti^{3+}, V^{2+}, Fe^{1+}, Fe^{3+}, Co^{1+}, Ni^{3+} (and presumably Ru^{1+} and Pd^{3+}) leaves no ambiguity about these products and ESR observation of Cr^{3+}, Fe^{2+}, Co^{2+} and Ni^{2+} removes ambiguity about the initial ions. The fact that X-irradiation of Ti-containing samples leads to the same spectrum as hydrogen reduction gives reasonable assurance of the initial presence of Ti^{4+}. The fact that hydrogen reduction shows the ESR spectrum or the characteristic R-line luminescence of V^{2+} is reasonable evidence that the precursor ion is V^{3+}†. The Cr^{2+} ion has been detected after H_2 treatment by ultrasonic resonance (Fletcher et al. 1966). Provided that the 4d ions all have the $4d^8$ configuration, as is now thought, viz. Ru^{1+}, Rh^{2+}, Pd^{3+}, the precursor ions should be the divalent forms.

The use of 5 eV irradiation is also known to produce V^{2+} and Fe^{3+}. The latter is said to be produced quantitatively from the divalent form by γ-irradiation after quench from a 5 min heating period in argon at 1375°K (Chen and Sibley 1967).

3.4. Luminescence properties of impurity ions

Characteristic luminescence phenomena are known for four substitutional impurity ions in MgO, viz. V^{2+}, Cr^{3+}, Mn^{4+} and Ni^{2+}. The single transition for the three $3d^3$ ions in octahedral symmetry corresponds to the pair of R lines from Cr^{3+} in Al_2O_3. Absorption in one of the d bands in the visible or near ultra-violet region is followed by a radiationless transition to the ^2E state; emission then occurs to the $^4A_{2g}$ ground state. The narrow line is accompanied by phonon sidebands at lower energy. Alternatively, the emission may be excited by X-irradiation. The transition occurs at $11\,498\,cm^{-1}$ for V^{2+} (Sturge 1963), $14\,323 \cdot 8\,cm^{-1}$ for Cr^{3+} (Deutschbein 1932, Schawlow 1962) and at $15\,279\,cm^{-1}$ for Mn^{4+} (Henderson and Hall 1967). The behaviour of the Cr^{3+} ion has been studied intensively. Its R line narrows markedly with decreasing temperature, reaching a line width of $0 \cdot 23\,cm^{-1}$ at $4 \cdot 2°K$ (Imbusch 1964). The line is split by application of a uniaxial stress along a principal axis; although both the $^4A_{2g}$ and the ^2E_g levels are split, the latter is split much more than the former (Imbusch et al. 1965). An independent measure of the ground-state splitting for a Cr^{3+} ion with a next-nearest-neighbour cation vacancy (Cr_T centre) is obtained from ESR measurements; this splitting is only $0 \cdot 16\,cm^{-1}$ (Wertz and Auzins 1957). At low concentrations of chromium, the most prominent subsidiary lines ('N' lines, Nebenlinien of Deutschbein) are a pair at $14\,317 \cdot 6$ and at $14\,204 \cdot 4\,cm^{-1}$ attributed also to Cr^{3+} associated with an nnn vacancy. A uniaxial stress applied along the defect axis causes an increase in the separation of these lines; hence the internal stress is a compressive one, with the Cr-vacancy distance less than the normal lattice constant of $0 \cdot 42$ nm. The variation with temperature of the relative intensities of the $14\,317 \cdot 6$ and the $14\,207 \cdot 4\,cm^{-1}$ lines according to the Boltzmann factor is evidence that the $110\,cm^{-1}$ splitting arises from the ^2E state. Another pair of lines at $14\,308$ and $14\,217\,cm^{-1}$ are weak at low Cr^{3+} concentrations, but their intensity increases rapidly with concentration. Again uniaxial stress experiments show that the symmetry axis is $\langle 100 \rangle$ or the equivalent. The relative intensities of these lines vary with temperature as the Boltzmann factor. The lines are attributed to the pair (or doubly associated) centre $Cr^{3+}\,O^{2-}\,\square O^{2-}\,Cr^{3+}$. An additional

---

† V^{3+} ions have been detected in non-cubic sites using APR (Rampton/1968).

splitting of 2 cm^{-1} is estimated on the basis of the relative intensity at 1·8°K and at 4·2°K of two lines resulting from the exchange interaction. Since the exchange interaction is so weak, it is satisfying to see the close correspondence between the splitting of the ^2E levels for the singly associated and double-associated centre; it is the same vacancy for both centres. Although the ESR spectrum of the chromium pair centre has not yet been described, that in which the position of the second Cr^{3+} ion is occupied by an Al^{3+} ion has been studied (Henderson and Hall 1967, Wertz and Auzins 1967). For this centre the zero-field splitting parameter D is slightly less than that for the singly associated centre, viz. 798·3 × 10^{-4} versus 819·2 × 10^{-4} cm^{-1}. This is suggestive of a very slightly increased Cr^{3+} vacancy distance in the double-associated centre. Careful examination of the luminescence spectrum in the region of the N-lines shows many satellites (fig. 14). Not all of these lines have been identified. However a careful comparison of the luminescence and ESR spectra (O'Donnell *et al.* 1976) reveal that the lines at 14 311·8 cm^{-1} and 14 214·8 cm^{-1} are due to the Cr^{3+} O^{2-} □O^{2-} Al^{3+} and the pair of lines at 14 302·5 and 14 227·4 due to the similar complex involving Ga^{3+}.

An additional Cr^{3+} emission line is reported to occur at 14 108 cm^{-1} (Glass 1967). Its intensity is said to be correlated with the Cr$_R$ centre which is associated

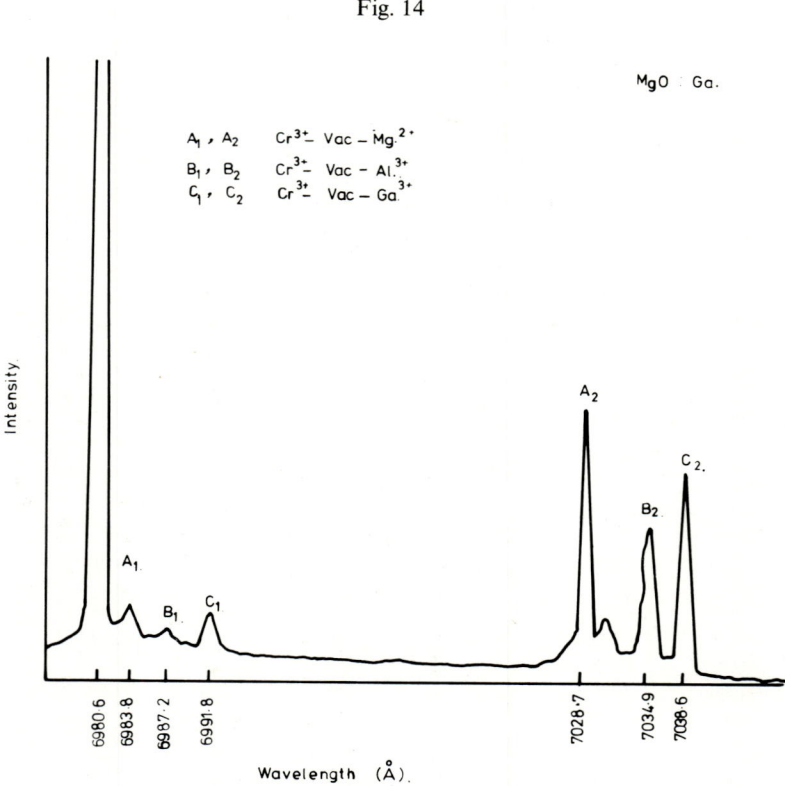

Fig. 14

Showing the fluorscence lines due to Cr^{3+} ions in tetragonal symmetry sites. Line positions given in text are from these recent measurements and may differ slightly from those quoted in the literature (after O'Donnell, Henry and Henderson 1976).

with a cation vacancy at the nn position ([011]-type direction). One would expect that there should be another line from the split ^2E levels. For this centre, the ground-state splitting is found from the ESR spectrum to be 0·8 cm^{-1} (Griffiths and Orton 1959). It is thus not surprising that an emission line from this centre is shifted to a much greater extent than the lines of either of the tetragonal Cr^{3+} centres.

Larkin *et al.* (1973) have made a very careful study of the absorption spectrum associated with the emission spectrum discussed above. They find the absorption coefficient in the R-line to be 2×10^{-20} cm, in excellent agreement with the theoretical value calculated assuming the transition to be pure magnetic dipole (Macfarlane 1970). The N-line absorption coefficient is larger, consistent with the mixed electric dipole and magnetic dipole processes involved in these transitions. These authors found no evidence in absorption of the 14,108 cm^{-1} transition, this failure throwing doubt on the suggestion by Glass (1967) that this transition is due to rhombic Cr^{3+}-vacancy centre. In fact a very recent study (Henry *et al.* 1976) shows that the ortho-rhombic chromium centre has the ^2E level lying above $^4T_1$: thus the luminescence is $^4T_1 \rightarrow {}^4A_2$ and polarized luminescence confirms this transition to emit at 9000 Å. Since the $^4T_1$ state is more sensitive to strain than ^2E, only a broad band luminescence is observed, the vibronic structure being broadened beyond resolution.

The 11 498 cm^{-1} emission line of V^{2+} in octahedral symmetry splits and shifts under uniaxial stress in a fashion analogous to that of Cr^{3+} (Sturge 1963). One might hope to find some evidence of emission lines from V^{2+} ions associated with a vacancy. Although this sounds unusual, it is to be noted that the normal valence state of most vanadium ions in MgO is V^{3+}; by reduction with hydrogen at high temperature or by irradiation with ultra-violet light or X-rays, one may get a large concentration of V^{2+} ions. For the ions which have been converted by ionizing radiation, one might expect that some were initially associated with a cation vacancy. While it is true that for the Cr^{3+} centres one notes a preferential change in valence state for the non-associated ions, there is also a small fractional change for the associated ions.

The cathodoluminescence spectrum (J. E. Ralph, unpublished) or the X-fluorescence spectrum of Mn^{4+} shows a number of lines in addition to the line at 15,279 cm^{-1} corresponding to octahedral symmetry. One at 15,168 cm^{-1} has been attributed to Mn^{4+} associated with a nnn cation vacancy (Glass and Searle 1967).

Occasionally, in undoped MgO crystals, one observes at 77°K or lower a set of four narrow emission lines at 20 880, 20 690, 20 490 and 20 310 cm^{-1} under X-ray excitation. Comparison with a lightly nickel-doped MgO crystal suggests that these are Ni^{2+} lines. A similar set of lines is observed in MgF_2 with low concentrations of Ni^{2+}, but at slightly lower energies due to a reduced crystal field (L. F. Johnson, private communication). The lines in MgO are observed to diminish in intensity with increased time of X-irradiation. This is in accord with the conversion of Ni^{2+} to Ni^{3+} under this treatment (Höchli *et al.* 1965). Upon standing, there is a recovery of the original fluorescence intensity, corresponding to a reversion to Ni^{2+}. It will be of interest to attempt to correlate these emission lines with the numerous sharp absorption lines which have been observed (Pappalardo *et al.* 1961).

Unambiguous association of thermoluminescence with an impurity ion in irradiation of MgO has been made for Cr^{3+} (Wertz *et al.* 1967). In some MgO crystals, especially those heated in O_2 at 1500°K, the most intense emission at about 390°K from a thermoluminescence peak induced by X-irradiation occurs near 14,300 cm^{-1}. A peak near this position has often been reported (Woods and Wright 1955, Saksena and Pant 1955, Gandy 1958, Hansler and Segelken 1960, J. E. Ralph, unpublished, Chen and Sibley 1967). Careful examination of the wave length distribution of this light reveals some indications of structure. Upon determining the spectrum of X-induced fluorescence in this same region at the same temperature, one finds similar evidence of structure. Progressive lowering of the sample temperature during X-irradiation shows the detailed correspondence of the partially resolved thermoluminescence lines with the R and N lines of Cr^{3+}. The unusual aspects of the lines in thermoluminescence is the requirement that a Cr^{3+} ion be generated in the excited ^2E state by a thermal process. Naturally, mere heating does not accomplish this. However, there are two possible mechanisms for generating the excited trivalent ion. These are (starting from the unexcited Cr^{3+} ion):

$$Cr^{3+} + e^- \rightarrow Cr^{2+}; \quad Cr^{2+} + h \rightarrow Cr^{3+*} \rightarrow R \text{ and } N \text{ emission,}$$

$$Cr^{3+} + h \rightarrow Cr^{4+}; \quad Cr^4 + e^- \rightarrow Cr^{3+*} \rightarrow R \text{ and } N \text{ emission.}$$

In ruby, the second process is believed to be the predominant mechanism, for one can observe the ESR spectrum of Cr^{4+} (Hoskins and van Steenwinkel 1964). Further, the orange colour of irradiated ruby, ascribed to Cr^{4+}, disappears as the thermoluminescence occurs (Compton et al. 1966). However, in X-irradiated MgO, microwave ultrasonic paramagnetic resonance measurements have shown the presence of Cr^{2+} ions (Fletcher et al. 1966); hence the second mechanism is also possible and perhaps is the more probable one in MgO.

Phosphorescence, cathodoluminescence or thermoluminescence peaks occurring in MgO subjected to ionizing radiation have been reported to occur in the range of 25 000–30 000 cm^{-1} (Bube and Stripp 1952, Woods and Wright 1955, Saksena and Pant 1955, Gandy 1958, Hansler and Segelken 1960, Chen and Sibley 1967, Wertz et al. 1967). Their position and intensity are highly dependent upon the sample and its thermal treatment. A luminescence peak at 29 000 cm^{-1} has been associated with photoconductivity, stimulable at 18 500 and 27 400 cm^{-1}. Subsequent to a thermoluminescence peak at about 390°K, the concentration of Fe^{1+} is found to be greatly reduced; a concomitant reduction in the concentration of Cr^{3+} in octahedral sites is highly dependent upon the sample (Wertz and Coffman 1965). It appears that some of the electrons released by Fe^{1+} are trapped by Cr^{3+} to form Cr^{2+}.

Several optical studies have been reported on the d^7 and d^8 ions. Typical of d^7 Co^{2+} has been investigated using mainly absorption techniques including magnetic circular dichroism. The absorption spectrum is extremely complex there being bands at 900 cm^{-1}, 16 000 cm^{-1}, 17 000 cm^{-1}, 20 000 cm^{-1}, 23 000 cm^{-1}, 25 000 cm^{-1}, 27 000 cm^{-1} and 28 000 cm^{-1} (Low 1958, Pappalardo et al. 1961 a, Ralph and Townsend 1968, Mann and Stephens 1972). Luminescence from Co^{2+} is best effected using the electron-excitation method (Ralph and Townsend 1968). most of the d–d transitions are exceptionally weak, being either allowed magnetic dipole or vibration-induced electric dipole in nature. The infra-red band (~ 900 cm^{-1}) has been identified as the $^4T_{1g} \rightarrow {}^4T_{2g}$ magnetic dipole transition. All bands in the visible – UV spectral region are vibrationally induced transitions, the fine-structure of which has been qualitatively interpreted as vibrational side bands of the no-phonon transitions. Quantitative band shape calculations, however, have not yet been performed on account of the calculative complexity introduced by the orbitally degenerate ground state of the Co^{2+} ions.

Pappalardo et al. (1961 b) first measured the optical absorption spectrum of Ni^{2+} ion in MgO and were able to construct a reasonably accurate energy level scheme. Ralph and Townsend (1970) measured both the optical absorption and the electron-beam excited fluorescence spectrum at 77 and 5°K. The observed transitions occur between the $^3A_2(t^6e^2)$ ground term and $^1T_2(t^5e^3)$, $^3T_1(t^5e^3)$, $^3T_2(t^5e^3)$ and $^3T_1(t^4e^4)$ excited terms. Comparable studies have been made by Manson (1971) and by Bird et al. (1972). Although the resolved structure associated with these transitions is faithfully reproduced by the various authors, there is some disagreement in the actual interpretation at least insofar as the influence of the Jahn–Teller effect on the $^1T_2$ band is concerned. The transition $^1T_2 \to {}^3A_2$ gives rise to a fluorescence band with structure centred near $\sim 20\,400\,\mathrm{cm}^{-1}$. This structure is similar to the set of four narrow emission lines at 20 880, 20 690, 20,490 and $20\,310\,\mathrm{cm}^{-1}$ observed occasionally at 77°K or lower in undoped MgO crystals under X-ray excitation. The lines in MgO are observed to diminish in intensity with increased time of X-irradiation. This is in accord with the conversion of Ni^{2+} and Ni^{3+} under this treatment (Höchli et al. 1965). Upon standing there is a recovery of the original fluorescence intensity, corresponding to a reversion to Ni^{2+}.

So far there have been no particularly systematic studies of transition or rare earth metal ions in CaO. However, Ratinen (1972) has recently surveyed the X-ray excited luminescence of rare earth ions, although no definite identification of the luminescence was attempted. More recently the fluorescence lines corresponding to the $^6P_{7/2} \to {}^8S$ transitions of Gd^{3+} in CaO have been reported by Hughes and Pells

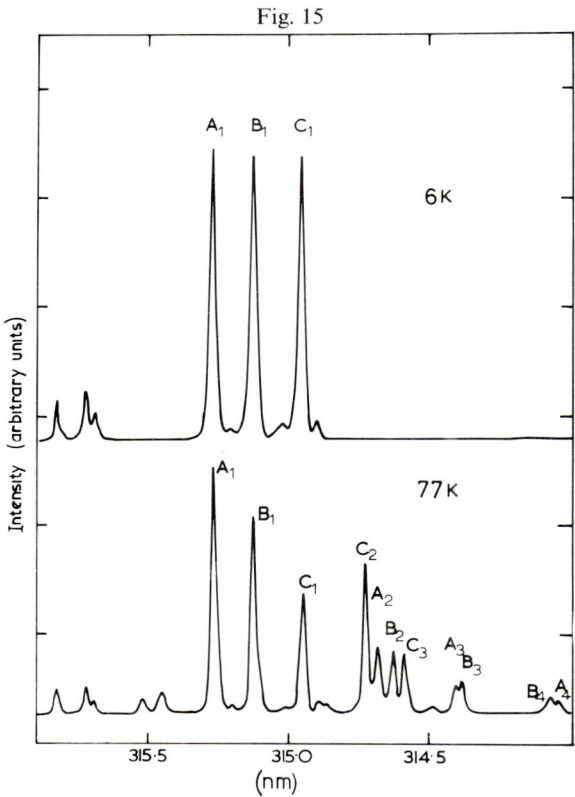

Fig. 15

Luminescence spectra due to Gd^{3+} ions in CaO crystals at 77°K and 6°K. The 3 groups of lines A, B and C refer to different site symmetries, of which A is known to be octohedral (after Hughes and Pells 1974).

(1974). At 77°K the three groups of lines near 32,260 cm^{-1} in fig. 15 are identified with Gd^{3+} in different site symmetries. Complementary EPR studies by McGeehin and Henderson (1974) confirm that the major symmetry sites are cubic and orthorhombic. The cubic centre has luminescence lines at 31,750, 31,723 and 31,786 cm^{-1}. The 31,732 cm^{-1} and associated lines are due to Gd^{3+} ions in orthorhombic symmetry, although even the EPR spectrum does not allow a complete specification of the environment of the ion. In this symmetry the EPR spectrum is characteristic of an $S = \frac{7}{2}$ ion with large values of D and E (McGeehin and Henderson (1974)).

§ 4. Trapped hole centres in oxide lattices

4.1. *Effects of ionizing radiation. V centres*

The several types of ionizing radiation—ultra-violet light, X-rays, γ-rays or electron beams of energies up to 2 MeV—give rise mainly to ionization rather than atomic displacement in the alkaline earth oxides. The locus of most primary events is doubltless the O^{2-} ion. Though it has a positive electron affinity in the gaseous state, the O^{2-} ion is stabilized to the extent of about 2 eV in the solid ionic state. If the behaviour of the alkaline earth oxides subsequent to the loss of one electron were to parallel that of the alkali halides, one would expect the formation of a X_2^- centre (self-trapped hole, Kanzig 1955). In this case an O$_2^{3-}$ molecule would be formed with its molecular axis along an [011]-type direction. Such a molecule would be isoelectronic with the well-known F$_2^-$ molecule. The O$_2^{3-}$ species has not as yet been detected. Its possible observation depends upon whether the relaxation of an O$^-$ ion towards an O^{2-} ion along an [011]-type direction leads to a lowering of the crystal energy.

The formation of large numbers of V type trapped-hole centres in MgO (Wertz *et al.* 1959) and CaO (Shuskus 1963) leaves little doubt that numerous free holes are released by irradiation and migrate to a negatively charged trapping site. For the V^--type centres the trapping site is an ever-present positive ion vacancy. Additional sites of hole trapping are ions which may change from the divalent to the trivalent state upon irradiation, including Fe^{2+} and Ni^{2+} (§ 3). The geometry of the defect proposed by Seitz to be responsible for a *V* band in the alkali halides is shown in fig. 16. This defect, though never observed in the alkali halides, is the important 'intrinsic' centre in the alkaline earth oxides. It will be shown to be associated with a characteristic violet colour of irradiated magnesium oxide. An apparent catalytic effect for *H–D* exchange in powders is attributed to V_1 centres.

The first evidence for this centre was obtained from its ESR spectrum in MgO (Wertz *et al.* 1959). At room temperature there are broad lines associated with a species which is unstable to thermal or optical bleaching. At 77°K these lines are about 30-fold narrower (0·4–0·6 G) and give the spectrum shown in fig. 17. (For the V_1 centre in CaO the line widths increase ten-fold in going from 77°K to 193°K.) For $H\|\langle 001\rangle$, the line at the $g_\perp$ position is twice as intense as the $g_\|$ line because other V_1 centres have $\langle 100\rangle$ or $\langle 010\rangle$ axes. The significant aspect of the spectrum is that it reflects tetragonal rather than octahedral symmetry. *A priori*, one might well have expected the hole to migrate rapidly among six presumably equivalent oxygen atoms about the vacancy. The existence of a tetragonal axis of symmetry requires that the hole, if not fully localized as in the crude representation of fig. 16,

Fig. 16

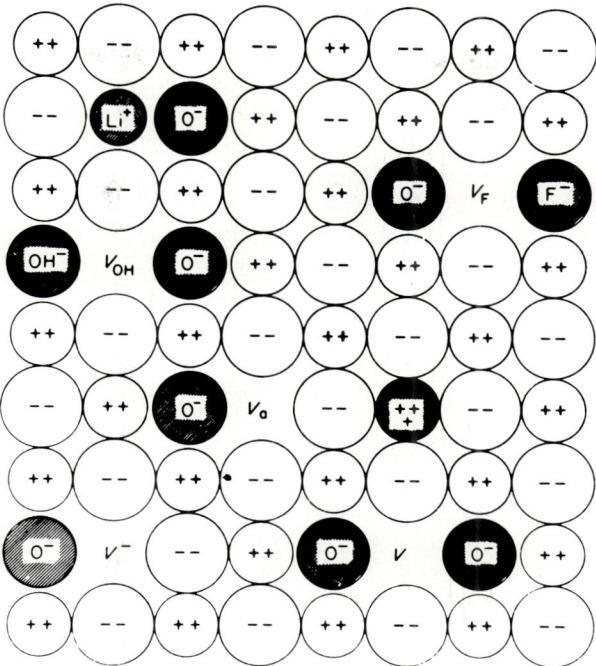

Models of trapped hole centres (after Hughes and Henderson 1972).

at least be preferentially localized along one axis. Appreciable localization would likely be accompanied by an extensive relaxation of the lattice.

The ESR spectrum shown in fig. 17 is quite complex due to interference from impurity centres. Lines labelled as V_F and V_{OH} were identified with the appropriate structures in fig. 16 by combined ESR/ENDOR studies. However, both V_F and V_{OH} spectra decay thermally much more rapidly than the V_1 spectrum, which may consequently still be studied with comparative ease. The ESR lines designated as V_1 have been much studied: they were for some years attributed to the intrinsic trapped

Fig. 17

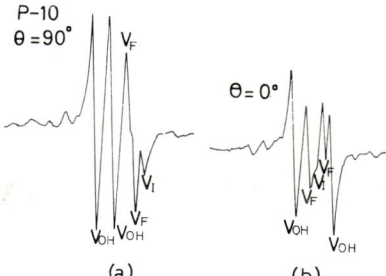

The ESR spectra of the V_1, V_F and V_{OH} centres in magnesium oxide with B_0 (a) perpendicular to the defect axis and (b) parallel to the defect axis.

hole centre, V^- (Wertz et al. 1959, Hughes and Henderson 1971). However, radiation damage and thermal stability data led Chen et al. (1969) to suspect that more than one defect exists with g-values identical to those of the V_1 spectrum. Subsequently a partial motional averaging of the V_1 spectrum was observed (Cowan, private communication, 1970), the significance of which was not fully realized.

The crucial test of the model for the defect associated with the V_1 spectrum remains the hyperfine structure. Early observations of hyperfine structure due to the ^{25}Mg isotope neighbouring the trapped hole were not interpreted fully (O'Mara et al. 1969, O'Mara 1969). The first ENDOR studies of the V_1 spectrum demonstrate the inherent weakness of assigning spectra to specific defects using only g-tensor data. Unruh, Chen and Abraham (1973) and DuVarney and Garrison (1973) reported that the ENDOR spectrum which results when the V_1 spectrum is saturated consists of five doublets due to ^{27}Al. This ENDOR spectrum is attributed to the charge neutral configuration V_a in fig. 16, the impurity being Al^{3+}. The evidence is compelling, especially when cognizance is taken of the observations conducted using many crystals of different origin as well as crystals doped with Al_2O_3 (Unruh et al. 1973). The stability of the V_{Al} centre varies with crystal: characteristically the half-life at room temperature is less than 10 hr. However in crystals irradiated with fast neutrons or 2 MeV electrons an ESR spectrum with identical g-values is observed, which does not yield the V_{Al} ENDOR spectrum. This spectrum appears to be due to the stable, intrinsic V^- centre. Unruh et al. (1973) note that this spectrum decayed very little during a 2-year period following prolonged electron irradiation. They point out that a negatively charged V^- centre should not decay by electron capture because of the Coulomb repulsion.

In ESR studies at 77°K using neutron irradiated MgO, Halliburton et al. (1973 a) have correctly identified and analysed hyperfine structure due to a ^{25}Mg isotope adjacent to the trapped hole on the axis of the defect. The usual sextet of hyperfine lines expected for a nucleus with $I = \frac{5}{2}$ is not observed, rather does the hyperfine structure show lines with varying intensities and spacings as well as extra lines. Halliburton et al. (1973 a) diagonalized the 12×12 matrix appropriate to the axial form of eqn. (11) using a computer. Detailed studies allow the authors to conclude that the spectra may be fitted to this equation using $a = +2.84$ MHz, $b = -0.86$ MHz and $P = +0.54$ MHz. Further the intensities may be accounted for only on the assumption that the hole be localized on one oxygen ion. This clearly contradicts the model for the V^- centre (Izen et al. 1973) which describes the ground state as a molecular orbital resulting from a linear combination of states centred on the two O^{2-} sites opposite each other across the vacancy.

Neglecting any hyperfine structure it is clear that the V_{Al} and V^- centres give identical ESR spectra at 77°K. As the temperature is increased the three lines of the anisotropic V_{Al} spectrum only broaden as a consequence of the temperature dependence of the ESR relaxation process. In contradistinction the V^- lines initially broaden on heating: between 215 and 230°K the $g_\parallel$ and $g_\perp$ lines begin to move towards one another, and at 245°K the lines coalesce to produce a single isotropic line with $g = \frac{1}{3}(g_\parallel + 2g_\perp) = 2.027$. This is shown in fig. 18, where both the V^- lines and the V_{Al} spectra are shown in the same crystal. The motional averaging of the V^- centre spectrum is readily apparent. Thus it is evident that the temperature dependence of the ESR spectra may distinguish between the V_{Al} and V^- centres.

The intrinsic V^- centre may also be produced in calcium oxide by neutron irradiation (Hall 1975) or irradiation with 2·MeV electrons (Abraham et al. 1975 a).

Fig. 18

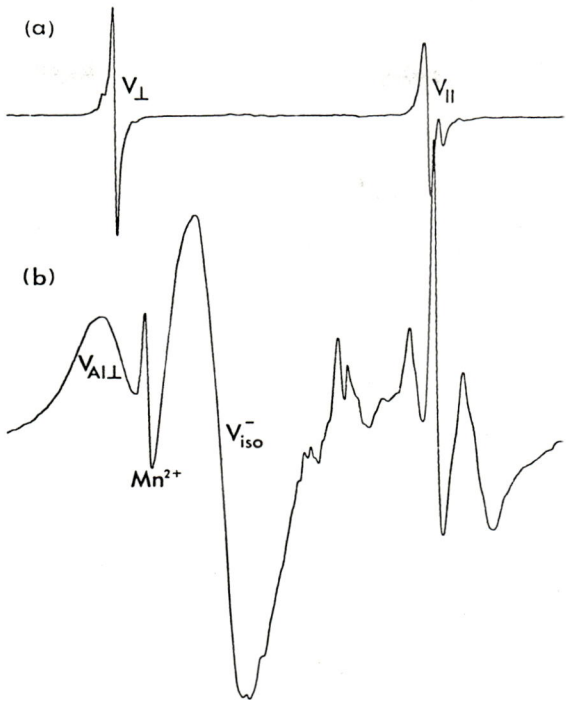

ESR spectrum of V_{Al} and V^- centres in MgO at (a) 77°K, where the V_{Al} and V^- spectra coincide due to identical g-values and (b) 290°K where the V_{Al} spectrum has broadened due to relaxation effects and V^- spectrum has coalesced to a single line at g_{iso} due to motional averaging (after Halliburton et al., 1973 b).

At 77°K the ESR spectrum has axial symmetry about the crystal $\langle 100 \rangle$ direction and averages to a single isotropic line above 195°K. Thus this behaviour is very similar to that observed for V^- centres in MgO. ENDOR studies on this spectrum show that there is no associated impurity (Abraham et al. 1975 a). The g-values are compared with those observed in MgO in table 7.

[Note many optical measurements were made prior to 1973: an optical band at photon energy $h\nu = 2\cdot 3$ eV was attributed to V_1 centres. Almost certainly this composite band resulted from the combined absorption due to V_{Al}, V_{OH}, V_F and V^- centres. Thus in the following section the term V_1 centre refers to such a *mélange* of centres, unless otherwise stated.]

4.2. Formation and stability of V-centres in MgO

The intensity of the V_1 spectrum which can be induced by irradiation is a function of thermal treatment. Treatments which give rise to an increased number of free positive vacancies should yield an increased number of V^- centres after irradiation. It was shown in § 1.3 that heating at a temperature of about 700°K leads to an increasing number of isolated vacancies. Hence, to obtain a high concentration of V^- centres, a period of heating above 1000°K should be followed by rapid quenching to prevent reassociation (Chen and Sibley 1967, Glass and Searle 1967, Henderson

Table 7. g-values and optical data for hole centres†

	$g_{\parallel}$	$g_{\perp}$	Δ (eV) obs.	Δ (eV) calc.†	Reference
MgO					
V^-	2·0032	2·0385	2·33	1·47	Wertz et al. (1963)
V_{OH}	2·0033	2·0396	2·19	1·29	Tohver et al. (1971)
[Li^0]	2·0043	2·0542	1·82	0·74	Henderson (1976)
[Na^0]	2·0055	2·0721	1·58	0·74	Abraham et al. (1972)
CaO					
V^-	2·0021	2·0697	1·91	1·00	Tohver et al. (1971)
V_{OH}	2·0018	2·0729	1·85	0·90	Henderson unpublished
[Li^0]	2·0019	2·0885	1·73	0·50	Tohver et al. (1971)
[Na^0]	2·0002	2·1234	1·39	0·50	Tohver et al. (1971)
[K^0]	2·0006	2·0962	1·6	0·50	Abraham et al. (1973)
SrO					
V^-	2·0010	2·0703	1·62	0·84	Henderson unpublished
[Li^0]	1·9999	2·0931	1·61	0·42	Henderson (1976)
[Na^0]	1·9957	2·1454	1·34	0·42	Abraham et al. (1972)
[K^0]	1·9966	2·1356	1·5	0·42	Abraham et al. (1973)

† No allowance has been made for distortion effects. Measurements reported at 4·2 K (ESR) and 77 K for optical peaks.

and Sibley 1971). Such a treatment of CaO yields V^- centres without subsequent irradiation (Tomlinson and Henderson 1969).

Necessarily, the maximum attainable V^--centre concentrations will be a function of the impurity content. In heavily doped MgO crystals no V^--centre spectrum is observed. Neither does one see the spectra of the various modified V centres of the next section. This is presumably because the impurity ions are a source of electrons which rapidly annihilate the holes. Alternatively, the impurities may themselves trap holes or else serve as electron hole recombination sites.

A stable trapping site is required for the electrons released from O^{2-} ions to form O^{1-} ions. The concentration of negative ion vacancies is usually undetectably small in these oxides (unless they have been subjected to neutron irradiation). Hence one requires impurity ions in a higher valence state to serve as electron traps, yielding an ion in a lower valence state. V^- centres are observed only in neutron-irradiated specimens subjected to small doses, e.g. 10^{17} nvt.

Once formed, the V^- centres may be bleached optically or thermally. Bleaching occurs with light over the entire visible spectrum, the bleaching efficiency reaching a broad maximum at $\sim 2·5$ eV. Hence it is desirable to enclose oxide samples in very thin aluminium foil during X- or γ-irradiation and for subsequent storage. When using a glass X-ray tube this is especially important, since the glow of the heater reduces the attainable V^--centre concentration. Increased yields of V_1 centres are obtained by irradiation at 77°K, since the half-life at room temperature may be as short as two hours (Chen and Sibley 1967). At 77°K the half-life is in excess of a week for the samples we have studied, provided they are kept in the dark. The decay is markedly sample-dependent and hence presumably is impurity-state dependent. Its decay does not follow at first-order law. A few V^- centres may persist after brief heating to 400°K. It is not yet established whether the decay proceeds via loss of the trapped hole from the vacancy region or else via the annihilation with an electron released from any of a variety of electron traps (Soshea et al. 1958). The metastable ions Fe^+ and Cr^{2+} would appear to be shallow traps and likely donors. The Fe^+ ion is known to give a large charge-release peak on heating to about 385°K (Wertz, unpublished).

Recently the Oak Ridge group have made a considerable effort to understand completely the formation and stability of the V type centres (Chen and Sibley 1967, Sibley and Chen 1967, Unruh et al. 1973 a, b, Abraham et al. 1974). The short-lived components of the optical V_1 band are due to V_{Al}, V_{OH}, and V_F which have lifetimes of the order of hours, whereas the V^- centre half-life is of order many months (Unruh et al. 1973, Chen and Sibley 1967, Chen, Kolopus and Sibley 1969). The nett effect is that at room temperature in the dark the total concentration of V_1 centres may have a half-life of days. If the temperature is raised the impurity-trapped hole associated centres decay rapidly by a hole release process. However the intrinsic V^- centre is not destroyed until temperatures in the range 600–850°K, whence OH^- ions thermally precipitate in the crystals absorbing Mg^{2+} vacancies in so doing (Henderson and Sibley 1971).

Abraham et al. (1974) have produced V^- centres using 2 MeV electrons, low energy X-rays and γ-rays. Their detailed studies demonstrate that V^- centres are formed by a process involving an essentially zero-energy threshold with no energy dependence. Furthermore the cross section for V^- centre production by 2 MeV electrons is large, 10^2–10^3 b, consistent with a displacement energy $\ll 5$ eV. (The measured O^{2-} ion displacement energy is ~ 60 eV.) They conclude that the dominant formation process for V^- centres involves ionization rather than the anticipated knock-on process. The formation of V^- centres at low doses of fast neutrons apparently results from the overwhelming γ-flux in most experimental reactors. Furthermore the process appears to be impurity dependent: V_{OH} (or V_{OH}^-) centres appear as precursors of V^- centres. The ionization causes the removal of H from V_{OH} rather than the displacement of the Mg^{2+} ion. This latter observation may offer an alternative explanation to that proposed above for the non-observation of V^- centres in crystals containing transition group elements cobalt, chromium and iron. Such crystals in general contain very low OH^- ion concentrations, as measured by infra-red spectroscopy. No similarly detailed studies have yet been reported of the production mechanism in CaO, or any other of the oxides.

4.3. *The nature of the trapped hole centre*

One describes the paramagnetic defect as a hole in a $2p_z$ orbital on an oxygen atom adjacent to the cation vacancy. That is, it is an O^{-1} ion in the ^2P ground state. It will be shown that the three-fold orbital degeneracy will be removed by a tetragonal field, leaving an orbital singlet (Kramers doublet) lowest. For such a centre, the basic states, which will be modified by spin–orbit coupling, are $|1, \pm\frac{1}{2}\rangle$, $|0, \pm\frac{1}{2}\rangle$ and $|-1, \pm\frac{1}{2}\rangle$. The effective Hamiltonian will contain an axial crystal field operator, spin–orbit coupling, and Zeeman terms. It may be written as:

$$\mathcal{H} = C[3L_z^2 - L(L+1)] + \lambda \vec{L} \cdot \vec{S} + \beta \vec{B} \cdot [\vec{L} + 2\vec{S}],$$
$$\mathcal{H} = C[3L_z^2 - 2] + \lambda[L_z S_z + \tfrac{1}{2}(\vec{L}_+ \vec{S}_- + \vec{L}_- \vec{S}_+)] + \beta \vec{B} \cdot [\vec{L} + 2\vec{S}]. \quad (36)$$

The fourth-order terms have been omitted from the crystal-field operator, since they will have non-zero matrix elements only for states with M_S values differing by four. If the magnetic field is taken as the z direction, the last term may be replaced by $\beta \vec{B}[\vec{L_z} + \vec{S_z}]$. The resulting 6×6 Hamiltonian matrix is factorable; the secular determinant yields the following three energies:

$$W_3 = C - \tfrac{1}{2}\lambda' + \frac{1}{2}\frac{\lambda'^2}{\Delta},$$
$$W_2 = C + \tfrac{1}{2}\lambda' \pm 2\beta B, \tag{37}$$
$$W_1 = -2C - \frac{1}{2}\frac{\lambda'^2}{\Delta} \pm \beta B.$$

These energies have been arrayed in the order in which they occur, i.e. the energy W_1 corresponds to the ground state. Here $\Delta = 3C$ and λ' has been taken as negative. Thus $g_{\parallel} = 2 \cdot 00$.

For $B \parallel x$, the last term in the Hamiltonian is written as

$$\tfrac{1}{2}\beta B[L_+ + L_- + 2(S_+ + S_-)].$$

The energies are most readily obtained from perturbation theory after making a new choice of ground-state wave functions to avoid an energy denominator of value zero. These functions are taken as $2^{-1/2}[|0, \tfrac{1}{2}\rangle + |0, -\tfrac{1}{2}\rangle]$ and $2^{-1/2}[|0, \tfrac{1}{2}\rangle - |0, -\tfrac{1}{2}\rangle]$. The resulting energies of the Kramers doublet constituting the ground state are given by:

$$W_{\pm} = -2C - \frac{1}{2}\frac{\lambda'^2}{\Delta} \pm \beta B \pm \frac{\lambda'\beta B}{\Delta}. \tag{38}$$

Hence

$$g_{\perp} = 2 \cdot 0023(1 - 2\lambda'/\Delta).$$

In these expressions Δ should have been replaced by $\Delta - \tfrac{1}{2}\lambda'$, and second order corrections $-[\lambda'/(\Delta - \tfrac{1}{2}\lambda')]^2$ added to the expressions for $g_{\parallel}$ and $g_{\perp}$. Generally the errors involved in the neglect of such terms are small (Hughes and Henderson 1971, Tohver et al. 1972), although they may be significant.

The splitting Δ between the p_x, p_y orbitals and the p_z orbitals has been calculated using a simplified crystal field model (Bartram et al. 1965, Hughes and Henderson 1971). The p_z orbital lies below p_x, p_y by an amount

$$\Delta = \frac{6}{5}\frac{e^2}{R^3}\langle r^2 \rangle \tag{39}$$

where $\langle r^2 \rangle$ denotes the integral for the radial part of the O^- $2p$ function. Bartram et al. (1965) find $\langle r^2 \rangle = 3 \cdot 04$ atomic units using a $2p$ wave function due to Hartree et al. (1939). In principle optical absorption due to transitions between the p_z ground state and the p_x, p_y orbitals can be induced only using light polarized perpendicular to the z-axis. However such transitions are expected to be strictly forbidden by the Laporte selection rule. Although more will be said about the electronic structure of trapped hole centres in §4.5 it is worthwhile to make some general comments about the V^- centres now. Values of Δ, $g_{\perp}$ and $g_{\parallel}$ are shown in table 7.

Examination of the experimental values $g_{\parallel} = 2\cdot0032$ and $g_{\perp} = 2\cdot0385$ for the V^- centre in MgO indicates that the former agrees well with prediction. This assurance that the sign of the splitting C is positive is in satisfying accord with expectation, for the $|0, \pm\frac{1}{2}\rangle$ states should represent a positive hole in a p orbital directed towards the effectively negative cation vacancy; this should correspond to the lowest energy. The magnetic resonance behaviour predicted for a negative value of C would have been totally different because the levels would have been inverted. Comparison of predicted and experimental factors requires an estimate of the value of λ for a free ion. A value of 135 cm^{-1} is obtained by an extrapolation of the λ values for the series Al^{4+}, Mg^{3+}, Na^{2+}, Ne^{1+}, $F°$ and O^{1-} (Bartram *et al.* 1965). Use of this value of λ with the optical transition energy $\Delta = 18\,300$ cm^{-1} gives $g_{\perp}$ (calc.) $= 2\cdot017$. Alternatively, the experimental value of $g_{\perp}$ leads to the experimental value $\lambda'/\Delta = -0\cdot018$ cm^{-1}. This is slightly more than twice the value ($-0\cdot0073$ cm^{-1}) found experimentally for the hole trapped at a cation vacancy in Al_2O_3 (Gamble *et al.* 1964). For this system also the $g_{\perp}$ value is larger than the calculated value (Bartram *et al.* 1965). The discrepancy in the $g_{\perp}$ value may arise from the model, since one should perhaps consider a small admixture of states having the positive hole in the $2p_x$ and $2p_y$ orbitals (Feuchtwang and Ferate, to be published). Since the absorption peak of the V^- centre in CaO occurs at 14,900 cm^{-1} we obtain $g_{\perp} = 2\cdot036$. As for MgO the quantity ($g_{\perp} - 2\cdot0023$) calculated is a factor of two smaller than is measured by EPR. The splitting Δ should be smaller in CaO, because of the reduced crystalline field, but only by a factor of d^{-3}, viz. 1·33. A theoretical study will be required to resolve this problem. We will have more to say about the nature of the optical band resulting from the V^- centres: before so doing we will discuss in some detail the properties of some of the modified V-centres.

4.4. *Modified V centres*

In § 4.1 we noted that positive holes may be trapped at sites involving impurity ions. These modified V centres are readily recognized by characteristic hyperfine structure in either the ESR or ENDOR spectrum. So far we have noted already the V_{Al}, V_{OH} and V_F centres, involving respectively Al^{3+}, OH^- and F^- ions as accidental impurities. In addition we will describe the properties of V_{OD} centres where OD^- ions have been deliberately added during arc fusion as well as the $[M^0]$ centres, holes trapped near unipositive ions such as Li^+, Na^+ or K^+.

4.4.1. *The V_{OH} centre*

Comparison of the V_1 ESR spectra in various MgO crystals reveals that the relative intensity of various satellite lines varies greatly. Some of these lines are markedly enhanced by heating at 700°K. All these lines have approximately the angular dependence behaviour and bleaching characteristics of the V_1 centre. Often the most prominent of the satellite lines is a doublet of equal intensity. The behaviour of these lines is given by the Hamiltonian:

$$\mathscr{H} = \beta[g_{\perp}(H_x S_x + H_y S_y) + g_{\parallel} H_z S_z]$$
$$+ b(S_x I_x + S_y I_y) + a S_z I_z - g_n \beta_n \mathbf{I} \cdot \mathbf{H}.$$

Here $S = \frac{1}{2}$, $I = \frac{1}{2}$, $g_{\parallel} = 2\cdot0033$ and $g_{\perp} = 2\cdot0396$. The values of the hyperfine parameters will be given later. From the ESR spectrum itself the hyperfine splitting appears to be *purely* anisotropic; with $H \parallel [111]$ the doublet separation vanishes, and hence the isotropic contribution is zero within experimental error. The maximum hyperfine splitting, occurring at the $g_\perp$ position, is 0.84 G. The angular dependence requires the nucleus responsible for the hyperfine splitting to lie along the axis of the centre. Since analyses show that fluorine is typically present in small quantities, it was surmised that the nucleus is ^{19}F. This proved to be incorrect; however, a V centre of this geometry (V_F centre) was identified later.

Electron-nuclear double resonance (ENDOR) experiments demonstrated unequivocally that the hyperfine splitting arises from hydrogen. They also permit a precise evaluation of the hyperfine coupling parameters as follows: $A/hc = 1.599 \times 10^{-4}$ cm^{-1}, $B/hc = 0.778 \times 10^{-4}$ cm^{-1}. Assuming purely spin–spin interaction, one may compute $\langle r^{-3} \rangle^{-1/3} = 0.322$ nm. Subtracting this value from the normal unit cell length of 0.420 nm, one obtains 0.098 nm as a rough measure of the spacing between the proton and the nearest oxygen atom. This is to be compared with the O–H spacing of 0.09 nm in $Mg(OH)_2$. Hence it is reasonable to assume that an OH ion substitutes for a normal O^{-2} ion, with the hydrogen atom lying next to the vacancy and along the axis defined by the O^- ion and the vacancy. This centre is therefore designated as V_{OH}. The configuration of this centre is shown in fig. 16.

From the isotropic component of the hyperfine splitting one computes $|\psi_s(0)|^2$ as 6×10^{18} cm^{-3}. This small magnitude and the angular dependence of the hyperfine splitting suggest a hole located primarily on a $2p$ orbital of oxygen. Experiments are currently in progress to interpret the hyperfine splitting of ^{25}Mg neighbours to map out more explicitly the wave function of the trapped hole.

It was noted early that MgO crystals which give a strong V_{OH} ESR signal also show a pronounced narrow infra-red band at 3296 cm^{-1}. The latter is always accompanied by a weaker band at 3310 cm^{-1}. All are more prominent in MgO crystals made from powders derived from $Mg(OH)_2$ than from $MgCO_3$ (Henderson, unpublished). Some MgO crystals show only a broad band in the 3300–3700 cm^{-1} O–H stretching region (Kirklin et al. 1965, J. E. Ralph, unpublished, Glass and Searle 1967). Upon heating at 700°K for some hours, the intensity of the broad band is diminished and well-defined 3296 and 3310 cm^{-1} lines appear. The intensity of the V_{OH} ESR signal is then greatly enhanced. If the infra-red absorption spectrum in this region is taken at 77°K after X-irradiation at the same temperature, the 3296 cm^{-1} band intensity is greatly reduced and a new band at 3323 cm^{-1} appears. The intensity of the latter corresponds closely to the intensity reduction of the former. Further, the intensity of the ESR V_{OH} signals from a number of different MgO crystals was found to parallel the intensity of the 3323 cm^{-1} band which could be induced by a fixed period of X-irradiation. Hence the 3323 cm^{-1} band is to be taken as the O–H stretching frequency in the V_{OH} centre. Upon heating an irradiated sample to room temperature, the V_{OH} spectrum is found to decay more rapidly than the V_1 spectrum. The exponential decay constant for the 3323 cm^{-1} band is found to be 6.1×10^{-5} sec^{-1} at 292°K. The activation energy for loss of the positive hole is 0.84 eV centre^{-1} (Dravnieks and Wertz, unpublished). Decay of the 3323 cm^{-1} band is accompanied by a corresponding regrowth of the 3296 cm^{-1} band. Hence the 3296 cm^{-1} band is to be assigned to an OH$^-$ ion with its hydrogen end adjacent to a positive ion vacancy. The increase of the O–H stretching frequency in the presence of the positive hole is another manifestation of the effect of a reduced crystalline electric field of a neutral doubly-associated vacancy (table 7). The occurrence of a weak band at 3323 cm^{-1} in crystals which have not been X-irradiated suggests the presence of a low concentration of stable V_{OH} centres. These are perhaps analogous to those V_1 centres which resist rather vigorous bleaching treatments. The enhancement of the 3323 cm^{-1} band in unirradiated but Li-doped MgO is interpreted as a result of the raising of the Fermi level by lithium (Glass and Searle 1967). The enhancement of both the 3296 cm^{-1} band and of the V_{OH} ESR signal after heating at 700°K and subsequent X-irradiation is a result of the increased association of OH$^-$ ions with positive ion vacancies in this temperature region. At higher temperatures, the OH$^-$ ions move to other sites. The especial interest and importance attached to the V_{OH} centre is that it has led to a description of the geometry of one stable site for hydrogen in an oxide.

In some MgO crystals one observes a strong infra-red band at 3700 cm^{-1}. This corresponds closely to the O–H frequency in Mg(OH)$_2$, and suggests the presence of small domains of Mg(OH)$_2$ (Cabannes–Ott 1956, Kirklin et al. 1965, Glass and Searle 1967). Confirmation of their existence has been obtained by electron microscopy of small precipitates which diminished in size in the electron beam. The electron diffraction pattern from these precipitates corresponded closely to the pattern from the hydroxide (Briggs and Bowen 1968) and showed that the precipitates grow with the c axis parallel to the MgO [111] axis and the a direction parallel to [110]. This orientation is identical with that deduced for the topotactic transformation of Mg(OH)$_2$ to MgO. A series of emission peaks near 1·4 eV have also been identified with Mg(OH)$_2$ precipitates (Sibley et al. 1968, Henderson and Sibley, 1971).

A broad infra-red band centred in the 3400 cm^{-1} region has been attributed to substitutional OH$^-$ ions in isolated sites. For these, it has been suggested that the hydrogen atom is directed interstitially (Glass and Searle 1967). The infra-red band at 3310 cm^{-1} does not change in intensity upon X-irradiation. Hence, it would seem reasonable to assign it to OH adjacent to a positive ion vacancy but with the V_{OH} geometry modified such that a hole cannot be stably trapped. It has been suggested that the association of a trivalent ion on the opposite side of the vacancy could give rise to the 3310 cm^{-1} line (Glass and Searle 1967).

Similar measurements have been reported for V_{OH} centres in CaO. The ESR and ENDOR spectra were first reported by O'Mara and Wertz (1970), the data essentially being very similar to the V_{OH} centres in MgO. However independent ESR measurements by Schirmer (1971) gave a discrepancy in the g-values of this centre. Subsequent ENDOR studies by Unruh et al. (1973) confirmed the g-values reported by Schirmer. Henderson et al. (1971) have reported the changes in the (OH$^-$) infra-red spectrum consequent upon γ-irradiation. Sharp lines 3418 cm^{-1} and 3447 cm^{-1} are shown to be due to a cation vacancy adjacent to OH$^-$ (V_{OH}^- centre) and V_{OH} centres respectively. The V_{OH} centre ESR and ENDOR spectra in SrO have also been studied (Blake et al. 1971).

An interesting variant on the V_{OH} centre involves the substitution of OD$^-$ for OH$^-$. This can be achieved by mixing D$_2$O with the appropriate powder (MgO or CaCO$_3$) prior to subjecting the powder to submerged arc fusion (Henderson and Sibley 1971, Abraham et al. 1975). NMR measurements on the MgO powders show that there is a very efficient $H \leftarrow D$ exchange (Henderson unpublished). ESR and ENDOR studies have been reported for both MgO and CaO (Henderson et al. 1971 b, Abraham et al. 1975). In CaO with the magnetic field parallel to the V_{OD} centre axis the ESR spectrum is a well resolved triplet confirming the involvement of a nucleus with spin $I = 1$. This structure was not observed in the isostructural MgO and ENDOR must be used. ENDOR spectra at three different orientations of the magnetic field relative to the defect axis are shown in fig. 19. The four lines arise

Fig. 19

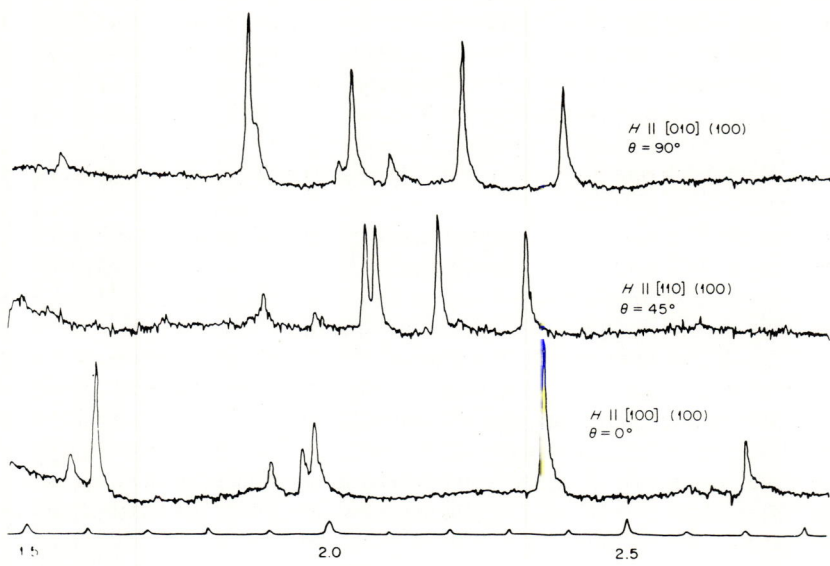

The ENDOR spectrum due to V_{OD} centres in MgO for three different orientations of the magnetic field in the (100) plane (after Henderson, Kolopus and Unruh 1971).

between the nuclear quantum levels $m_I = 0$ and $m_I = \pm 1$, two each for the electronic levels $m_S = \pm \frac{1}{2}$. Some weaker lines due to ^{25}Mg are also observed.

4.4.2. *The V_F centre*

The V_F centre in MgO, shown in fig. 16, is an analogue of the V_{OH} centre in two respects. First, both ions have a single negative charge, and therefore they are seen by the lattice as the sites of positive charge. Secondly, both occupy an oxygen site along the hole–vacancy axis; hence the fluorine nucleus, like the proton, gives rise to a hyperfine doublet by interacting with an unpaired electron in a $2p_z$ orbital on the O^- atom. The ESR spectrum in fig. 17 also shows a pair of doublets for $H\|[001]$ and $H\|[100]$. For the V_F centre, $g_\| = 2.0031$ and $g_\perp = 2.0388$; the hyperfine splitting has the same principal axes as the g components and is given by $0.08 + 0.33(3\cos^2\theta - 1)$G. While the value of the isotropic splitting seems small, it is about five times that (0.016 G) for the V_{OH} spectrum. However, the anisotropic splitting is several times less than that of V_{OH}. Thus the amount of polarization transferred to the p orbitals of fluorine is exceedingly small. Again, the trapped hole appears to be strongly localized.

Rather better resolution of the hyperfine structure is achieved by resort to ENDOR (O'Mara *et al.* 1969). ESR and ENDOR studies of the analogous centre in CaO give very similar results to those in MgO (Schirmer 1971, Unruh *et al.* 1973).

It should be noted that for these centres, V_{OH}, V_{OD} and V_F, the isotropic hyperfine interaction is almost zero and in consequence much smaller than the anisotropy in the interaction. This is a clear case where the effective field space quantizes the nuclear field rather than the applied field. Necessary corrections have been made which take this into account in studies of the angular dependence of the ENDOR spectrum of V_{OH}, V_{OD} and V_F in MgO. Alternatively the hyperfine constants are obtained from the $\theta = 0°$ or $90°$ orientations (when no quadrupole term is involved), or by a least squares fit of the data to the spin Hamiltonian.

4.4.3. *Other centres*

The V_a centre, for which the presumed structure is shown in fig. 16, was mentioned in § 3 as an example of a doubly associated centre. It appears to have the largest $g_\perp$ value (2·0390) of any of the V-type centres in MgO. Its signal intensity is usually weak. If one is able to prepare a sample in which it is sufficiently intense, it may be possible to do an ENDOR experiment to verify its structure and identify the associated impurity ion.

It is obvious that one trapped hole only half compensates the effective charge of a positive ion vacancy. Recalling the doubly associated centres of table 3 one anticipates the possibility of trapping a second hole at a V^- centre. There appears to be no counterpart for this defect, which will be called the V^0 centre. Electrostatic repulsion will favour the collinear orientation of the two trapped holes on opposite sides of the positive ion vacancy. For $H\|[001]$, the ESR spectrum seen upon X-irradiation and observed at 77°K consists of pairs of lines approximately centred upon the lines of the V^- centre (Wertz et al. 1959). Spin–spin interaction in this triplet state system leads to a zero-field splitting of the usual form.

The spin Hamiltonian which describes the spectrum is:

$$\mathcal{H} = \beta g_\perp(\hat{S}_x B_x + \hat{S}_y B_y) + g_\| \hat{S}_z B_z + D \left[\hat{S}_z^2 - \frac{S(S+1)}{3} \right] \quad (40)$$

with $S = 1$, $g_\| = 2·0032$, $g_\perp = 2·0408$ and $D/hc = -212·5 \times 10^{-4}$ cm^{-1} (Wertz et al. 1959, Wertz, unpublished). Measurements of the relative intensities of the ESR lines at low temperatures indicates that the state $|-1\rangle$ lies lowest; hence the sign of D is negative. If it is assumed that D is wholly attributable to spin–spin interaction, a calculation of the mean separation of the two holes gives the value 0·49 nm. This is large compared with the normal lattice spacing of 0·420 nm, though some extensive relaxation is expected. Observations of the analogous centre in CaO are fully in accord with the model discussed for MgO (Henderson and Tomlinson 1968, Abraham et al. 1975 a).

Decay of this centre is expected to proceed via production of the V_1 centre in both MgO and CaO, i.e.

$$V^- + h \rightleftharpoons V^0.$$

The binding energy of the second hole is expected to be less than 1 eV, since the defect is far less stable than the V^- centre. The dissociation of the V^0 centre occurs in MgO and CaO near 250°K and results in an increase in V^- centre concentration. If the mobile hole is trapped by an isolated cation vacancy a second V^- centre is formed. Thus the sum $2[V^0] + [V^-]$ will be conserved, assuming this to be the only process. This appears to be the case in CaO (Abraham et al. 1975 a). In MgO the hole appears to migrate to other traps since the sum $[V^0] + [V^-]$ appears to be conserved, indicating that only a single V^- centre is formed for every decaying V^0 centre (Chen et al. 1975).

The isolated cation vacancy, $V^=$ centre is not the only site in the oxide lattice at which charge compensation might occur. Wertz and Auzins (1959) suggested that a unipositive ion, M^+, may charge compensate locally for trivalent ions such as Cr^{3+}, thereby lowering the site symmetry. Schirmer (1968, 1971) carried over this idea into the defect area in his work on ZnO and BeO, and later on MgO, CaO and SrO: the structure of $[M^0]$ centres is shown in fig. 16. The ion initially used was Li^+, substituent for the appropriate divalent host ion. X-irradiation produced positive holes trapped on O^{2-} ions adjacent to Li^+—in the nomenclature of Sonder and Sibley (1971) the $[Li^0]$ centre. The hyperfine structure immediately identifies the defect since the nuclear spin $I = \frac{3}{2}$. Subsequently workers at Oak Ridge extended this work quite considerably with detailed investigations of the magnetic resonance and optical properties of $[Li^0]$, $[Na^0]$ and $[K^0]$ in MgO, CaO and SrO (Tohver et al. 1972, Abraham et al. 1972, 1973). In some cases motional averaging was observed in the EPR spectrum. This is made clearer in the ENDOR studies on these centres (Rius et al. 1970, Rius and Hervé 1974, Abraham et al. 1975): surprisingly the hopping between the six equivalent O^{2-} sites occurs even at $4 \cdot 2 °K$. This motion distinguishes between the centre-symmetric centres V^-, $[M^0]$ on the one hand and those such as V_{OH}, V_F etc. where the impurity ion adjacent to the vacancy determines and fixes the position of the trapped hole.

Finally we comment on two trapped hole centres whose structures are akin to the molecular ion species observed in the halide lattices. Rius and Cox (1968) have identified the mixed molecular ion OF^{2-} in MgO irradiated with neutrons at liquid neon temperatures ($27 °K$). The molecular ion axis is parallel to a crystal [111] axis, the OF^{2-} replacing a single O^{2-} ion: as such it is an interstitial centre. Hall (1975) has observed two different defects with $S = \frac{1}{2}$ which are tentatively identified as interstitial O_2^- ions aligned (i) along $\langle 110 \rangle$ (ii) with axes tilted $14 \cdot 5°$ from $\langle 110 \rangle$ lying in the corresponding (001) plane.

4.5. *Optical absorption of the V^- centre*

The optical absorption band at $2 \cdot 3$ eV which is now attributed to the V centres was first detected in MgO crystals which had been subjected to ultra-violet irradiation. Since the intensity of the violet coloration varied markedly from specimen to specimen, it was concluded, reasonably enough, that it is due to an impurity (Hibben 1937). Parallel optical and ESR experiments have established the identification of the optical band (Wertz et al. 1963, Tohver et al. 1972, Chen and Sibley 1967). Its half width at $77 °K$ is $0 \cdot 96$ eV, while at $295 °K$ it is $1 \cdot 09$ eV. The low-energy side appears to be accurately gaussian. The oscillator strength has been estimated as $0 \cdot 1$ to $0 \cdot 2$; however, this is probably only a lower limit (Chen and Sibley 1967).

The $2 \cdot 3$ eV band of the V_1 centre is overlapped on the high-energy side by numerous bands. Besides two at $4 \cdot 3$ and at $5 \cdot 7$ eV due to Fe^{+3} (Soshea et al. 1958, Peria 1958), there are bands at $3 \cdot 1$ and $3 \cdot 7$ eV. In most instances the latter bands are lower in intensity than that at $2 \cdot 3$ eV. However, it is possible to cause the $3 \cdot 1$ eV band to have a greater amplitude than that at $2 \cdot 3$ eV by heating to $1500 °K$ and quenching. The $3 \cdot 1$ eV band, along with other weaker ones, is responsible for the brown coloration shown by some MgO crystals initially and by all after the heating and quenching treatment. These bands have not yet been assigned to specific defects. They may be due to diamagnetic species, since no correlation with ESR spectra has yet been possible (Dravnieks and Wertz, unpublished).

The existence of a centre with a tetragonal axis and a known optical band immediately suggests experiments on polarized excitation or bleaching. In the alkali halides, it is possible to cause reorientation of the axes of X_2^- centres while exciting along their axes with polarized light in the X_2^- band (Delbecq *et al.* 1958). It is then possible to demonstrate occurrence of optical dichroism. However, attempts to do polarized excitation or polarized bleaching of the V_1 centres in MgO at 20°K or at 4·2°K have failed (Chen and Sibley 1967, Wertz, unpublished). Success in such experiments would require localization of a hole for a time at least of the order of seconds. The ESR spectra indicate only that because the lines are narrow, the lifetime must be long compared with the inverse line width, i.e. of the order of more than 10^{-7} sec. Reorientation about the vacancy doubtless occurs rapidly as compared with the optical time scale.

Of course as we mentioned in § 4.1, the term V_1 centre used here implies the presence of V^-, V_{OH}, V_{Al} and possible V_F centres. Certainly in the earlier work no account was taken of this since the subtle nuances of the impurity problem so beautifully illustrated by the ENDOR work were not yet known. Thus in general one equated V_1 with V^- and attributed to this centre in MgO a peak at 2·3 eV, with half-width of $\sim 1·0$ eV and large oscillator strength. The understanding which we now have of the impurity effects suggests that one should try to isolate the optical bands of the specific centres. However, no matter what change in the environment the optical transition starts out from a well-localized p_z orbital on the O^- ion. The model presented in § 4.3 assumes $p_z \to p_x$, p_y transitions and predicts a splitting between such states in terms of eqn. (39). The constant multiplicative factor $6e^2/5R^3$ arises from that part of the crystal field potential resulting from the cation vacancy (Hughes and Henderson 1972). This vacancy produces a contribution to the crystalline electric field of

$$C = -\frac{2e}{R_0^3}$$

where R_0 is the anion-cation separation. In the case of the $[M^0]$ centres the factor 2 is obviously missing: the implication is that the value of Δ is reduced by a factor 2 for $[M^0]$ centres relative to V^- centres. However the effect of Al^{3+} or OH^- ions requires us to add terms of $+e/(3R_0)^3$ or $e/(2R_0)^2$ respectively to C. In these cases the shifts are of order $(0·03–0·13) \times \Delta$, which bearing in mind the band width, will be difficult to detect. One further effect cannot be neglected, the more complex problem theoretically of the lattice relaxation around the defect. Neglecting the details of the relaxation we replace R_0 by $R_0(1+\varepsilon)$, where $\varepsilon = \delta R/R_0$ scales the change in interionic separation. For small elastic distortions, δR, then the distortion effect introduces a shift $\pm 3\varepsilon\Delta$, the sign depending upon whether there is a contraction or expansion. Typically the estimated values of ε lie in the range $+(3–7)\%$ giving rise to shifts to smaller energies of order 9–21%. Obviously great care has to be exercised in isolating the various contributions.

Tohver *et al.* (1972) and Kappers *et al.* (1972) have independently separated the V_{OH} and V^- bands as peaks at 2·19 eV and 2·33 eV respectively at room temperature. The shift is about 6%, rather less than is predicted on the simple crystal field model: it is however in the correct sense. The shifts in peak position of the $[Li^0]$, $[Na^0]$ and $[K^0]$ centres are also in the correct sense. This is shown clearly in table 7. However comparison of all the results shows that the simple crystal field model seriously underestimates the band peaks. This is disappointing albeit not necessarily surprising. Unfortunately it cannot be regarded as the only puzzling feature of the optical properties.

Comparison of the relative intensities of the optical bands and ESR intensities for all these centres, V^-, V_{OH} and $[M^0]$, shows the oscillator strength to be $f \simeq 0.1$, a value more typical of an allowed transition. The transition considered here between an orbital singlet $A(p_z)$ state and an excited orbital doublet $E(p_x, p_y)$ state derived from the same ^2P term of the O^- ion is strongly forbidden by the Laporte selection rule. The oscillator strength, f, should be not larger than about 10^{-5} even though the C_{4v} crystal field removes the parity violation of the electric dipole transition. Nor should odd parity phonons supply such a large oscillator strength.

Given the known localized ground state perhaps one should concentrate on the nature of the excited state. The sensitivity of the magnetic circular dichroism (MCD) of the V^- band in MgO to the ground state EPR leaves no doubt about the association of the band with V^- centres (Izen et al. 1973). Bowler and Searle (1971) find that the V_{Al}/V^- centre has a rather diffuse excited state of radius ~ 4 nm, in agreement with the MCD data of Izen et al. (1973). However to account for the asymmetric MCD these workers invoked that even in the ground state the hole is shared by two O^{2-} ions on opposite sides of the vacancy; various modes of possible delocalization are allowed in the excited state. Experiments which particularly relate to this problem delocalized levels involve the application of an electric field to crystals containing V^- centres (Rose and Cowan 1974, Rius and Hervé 1975, Henderson 1976). The temperature dependence of the induced dichroism in the V^- band in MgO can only be interpreted in terms of a reorientation of the localized ground state and not due to excited state splittings. These studies yield an effective dipole moment of $\mu = 3.4 \pm 0.4$ eÅ; the centres however must still freely rotate even at $4.2°$K. In fact the low temperature data are particularly important since the electric polarization (which follows a Brillouin-like functional T dependence) should either saturate or go to zero if the centres become locked at a particular site. That neither situation obtains is interpreted in terms of the freely orienting electric dipoles (V^- centres) being responsive also to the relatively large, random electric fields due to other charged defects in the crystal. Rose and Cowan (1974) assume that such charged defects are inevitably point defects: in ionic solids dislocation lines also are charged and may contribute to the internal electric field. The dislocation substructure is simplified by annealing at 2200°K followed by slow cooling to 1000°K and quenching to room temperature. The electric dipole moment measured then is only $\mu = 3.1 \pm 0.3$ eÅ and the maximum polarization is 0.85: this latter compares with the value of 0.71 reported by Rose and Cowan (1974). Investigation of the optical dichroism as a function of wave length shows a clear splitting of the excited states, which is interpreted as a delocalized molecular orbital. Furthermore the dichroism shows an $E(p_x, p_y)$-like state at ~ 1.5 eV, very close to the theoretically predicted position using the simple localized orbital model (Bartram et al. 1965, Hughes and Henderson 1972).

There is impressive confirmation of these data in the rather beautiful EPR studies of V^- and $[Li^0]$ centres in MgO under either uniaxial stress or external electric field (Rius et al. 1976). Both types of perturbation produce realignment of the O^- ions as a consequence of the applied perturbation. From electric field dependence studies at 77°K they determine the electric dipole moment to be $p = 3.2 \pm 0.1$ eÅ, in impressive accord with the optical data. From reorientation under uniaxial stress they find the elastic dipole moment of V^- centres to be $(9.6 \pm 0.4) \times 10^{-3}$ K kgf^{-1} cm^2. Furthermore Rius et al. have measured the reorientation time at low temperature using the ELDOR technique: they conclude that the dipole rotation process is a one phonon assisted tunnelling process.

This type of measurement is rather more difficult to make using CaO crystals owing to problems with hydroxide surface layers. However measurements on both V^- and $[Li^0]$ centres agree with the findings on MgO. The measured dipole moments are respectively 4.2 ± 0.2 eÅ and 1.6 ± 0.4 eÅ for V^- and $[Li^0]$ centres.

4.6. *More on the nature of trapped hole centres*

The discussion so far has concentrated on the structure of trapped hole centres rather than their electronic structure. However detailed knowledge of g-shifts, hyperfine interactions and optical transition energies give a considerable body of meaningful information on the electronic structure. We have seen that the optical absorption spectrum almost certainly involves on the ESR time scale a strongly localized orbital singlet state, $A(p_z)$, and a rather diffuse excited level of undetermined character. It is therefore appropriate to discuss only the ground state of these centres in detail. A point to be noted is that the observed g-shifts are quite extensive for all the centres studied (V^-, V_{Al}, V_{OH}, V_F and $[M^0]$). The origin of these shifts is the admixture of angular momentum into the ground state singlet via the spin-orbit coupling interaction (§ 4.2). However for a hole centre having appreciable orbital momentum which interacts with a ligand nuclear spin one must consider additional terms in the hyperfine Hamiltonian. Such terms involving dipole–dipole interaction between a nuclear magnet and the electron orbital moment were first calculated for X_2^- centres in the alkali halides (Schoemaker 1966 and 1973, Jette 1969); applied to axial trapped hole centres these additional terms take the form (Schirmer 1971)

$$\mathcal{H}(I, S) = \tfrac{1}{2}\Delta g_\perp B(I_x S_x + I_y S_y). \tag{40}$$

Usually in ESR the hyperfine structure is reported in terms of the measured parameters $A_\parallel$ and $A_\perp$. These are then related to the isotropic (a_0) and anisotropic (b) interactions through:

$$\begin{aligned} A_\perp &= a - b(1 + \Delta g/2) \\ A_\parallel &= a + 2b \end{aligned} \tag{41}$$

in units of $h/g\mu_B$ (see § 2.1.1). The correction $\tfrac{1}{2}\Delta g b$ to $A_\perp$ arising out of $\mathcal{H}(I, S)$ is unimportant for V^- centres since a and b are relatively large. However for V_F, V_{OD} and V_{OH} this correction is necessary in order that a consistent trend is obtained through the series MgO to SrO (table 8). If the additional term is neglected then the signs of a and b are observed to be the same in MgO; in CaO and SrO the signs are opposite. As first noted by Henderson and Garrison (1973) this sign change disappears when eqn. (41) is used, a and b both being positive. One should bear in mind that the hole is localized on an O^{2-} ion one lattice repeat distance from the OH^- or F^- ion. In the V_{Al} centre the Al^{3+} ion is a further half repeat distance from the paramagnetic species. In each case a is small requiring that the positive hole be well localized on the O^{2-} ion. Thus we can treat the anisotropic constant as arising from dipolar interaction between point dipoles: using eqn. (17) we can estimate the average distance R between the O^- ion and the interacting nucleus. Correcting eqn. (17) for the finite extent of the O^- ion, gives hb in megahertz as:

$$hb = \frac{g\beta g_N \beta_N}{R^3}\left(1 + \frac{12\langle\rho^2\rangle}{5R^2}\right)$$

Table 8. Hyperfine structure parameters for trapped hole centres

Centre	Host	Nucleus	$A_\parallel$ (MHz)	$A_\perp$ (MHz)	a (MHz)	b (MHz)	hP (MHz)	Reference
V^-	MgO	^{25}Mg	+1·12	+3·70	+2·84	−0·86	−0·54	Halliburton et al. (1973)
V_{Al}	MgO	^{27}Al	±0·178	∓0·074	±0·030	±0·790	∓0·549	DuVarney and Garrison (1973); Unruh et al. (1973)
V_{OH}	MgO	^{1}H	±4·843	∓2·315	±0·101	±2·371	—	Unruh et al. (1973)
	CaO	^{1}H	±2·758	∓1·386	±0·028	±1·365	—	Unruh et al. (1973)
	SrO	^{1}H	±2·032	∓1·033	±0·013	±1·009	—	Blake et al. (1971)
V_{OD}	MgO	^{2}H	±0·734	∓0·356	±0·012	±0·361	0·173	Henderson et al. (1971)
	CaO	^{2}H	+0·423	−0·212	+0·005	+0·209	0·188	Abraham et al. (1975)
V_F	MgO	^{19}F	±2·1072	∓0·6894	±0·234	±0·923	—	O'Mara et al. (1968)
	CaO	^{19}F	+1·553	−0·461	±0·226	±0·664	—	Unruh et al. (1973)
$[Li]^\circ$	MgO	^{7}Li	∓0·56	∓6·96	∓4·80	±2·12	—	Schirmer (1971)
	CaO	^{7}Li	±0·006	∓3·87	∓2·49	±1·34	∓0·004	Tohver et al. (1972)
	SrO	^{7}Li	±0·03	∓2·8	∓1·8	±0·9	—	Schirmer (1971)
$[Na]^\circ$	MgO	^{23}Na	∓3·02	∓11·4	∓8·61	±2·79	~±0·8	Abraham et al. (1972)
	CaO	^{23}Na	±5·40	±11·10	∓9·14	±1·84	±1·8	Tohver et al. (1972)
	SrO	^{23}Na	±5·04	∓7·72	∓6·83	±0·89	∓1·3	Abraham et al. (1972)
$[K]^\circ$	CaO	^{39}K	8·99	—	—	—	—	Abraham et al. (1973)
	SrO	^{39}K	3·39	—	—	—	—	Abraham et al. (1973)

R being the true O^--nuclear distance and ρ the mean squared radius of the trapped hole. Rius and Hervé (1974) find a value of $\langle\rho^2\rangle = 0.73$ Å^2 from ENDOR data in BeO. For the centres of interest here $R \approx 4.0$ Å and in consequence the correction is only of order 3%. If we write $R = R_0(1 + \varepsilon)$ we can estimate the distortion ε parallel to the defect axis since

$$hb \simeq g\beta g_N\beta_N \left(1 + \frac{12\langle\rho^2\rangle\{1 - 2\varepsilon\}}{5R_0^2}\right)\left(\frac{1 - 3\varepsilon}{R_0^3}\right). \quad (42)$$

This analysis (see table 9) shows that for all V_{OH}, V_{OD} and F_F centres the average distortion around the centre is typically 3–6% outwards.

Since b is positive then so also is a. From a we determine $|\psi(0)|^2$ the hole density at the impurity nucleus. As expected this is a relatively small quantity, showing that

Table 9. Interaction lengths and $|\psi(0)|^2$ values for trapped hole centres.

Host	Centre	ε	R(Å)†	$\psi(0)^2$ 10^{20} cm^{-3}
MgO	V_{OH}	0·033	4·356	1·52
CaO	V_{OH}	0·035	4·985	0·42
SrO	V_{OH}	0·019	5·08	0·20
MgO	V_{OD}	0·037	4·374	1·16
CaO	V_{OD}	0·037	4·997	0·45
MgO	V_F	0·054	4·443	3·76
CaO	V_F	0·025	4·956	3·63
MgO	V_{Al}			
MgO	[Li^0]	0·25	2·627	$-186·7$
CaO	[Li^0]	0·25	3·012	$-96·9$
SrO	[Li^0]	0·34	3·398	$-70·0$
MgO	[Na^0]	0·031	2·173	$-492·1$
CaO	[Na^0]	0·017	2·448	$-522·4$
SrO	[Na^0]	0·196	3·038	$-390·4$

† The lattice spacing is assumed to be $R_0 = 4·2114$ Å for MgO (Henderson and Bowen, 1971), and 4·813 Å for CaO, 5·08 Å for SrO. Note to obtain R for V_{OH} and V_{OD} centres the bond length for OH$^-$ and OD$^-$ ions has been added to the OH$^-$ – O$^-$ ion separation.

the hole wave function has only a small overlap on the ion diametrically disposed across the cation vacancy. Furthermore for a particular kind of centre $|\psi(0)|^2$ scales in a rather obvious way with respect to the calculated value of R.

This type of analysis is easily extended to the [M^0] centres with the results also shown in tables 8 and 9. Note that the Fermi contact interaction is relatively weak; consequently $|\psi(0)|^2$ is small. Thus one can reasonably treat any anisotropy as due to the mutual dipolar interaction between electron and nuclear spin. The distortion parameter ε then assumes large values, in excess of 20%, for the [Li^0] centres which bearing in mind the result for [Na^0] centres appears to be mainly due to the smallness of Li$^+$. Alternatively we may appeal to the quadrupole interaction and extract information from the electric field gradient using the techniques developed by Feuchtwang (1962). Accordingly the measured electric field gradient is written as a sum of terms due to the distortion and polarization of the lattice, the effective charge of the defect in the absence of the trapped hole and the distribution of the trapped hole over the defect site. Thus the effective field gradient, eq, is written in terms of

$$q = q_P(1 - \gamma_\infty) - \left[\frac{b}{g\beta g_N \beta_N} - \frac{4}{R^3}\right] - \frac{2\gamma_\infty}{R^3} \qquad (43)$$

which is obtained from the quadrupole interaction hP (in MHz) since

$$hP = \frac{3e^2 Qq}{4I(2I-1)}$$

In these equations Q is the quadrupole moment of the particular nucleus, q_P is the contribution to the field gradient produced by the polarization of the surrounding lattice and γ_∞ is the Sternheimer antishielding factor. Now $b/g_N \beta \beta_N \approx g\langle R^{-3}\rangle$ so that eqn. (42) simplifies to

$$q = \left(q_P + \frac{2}{R^3}\right)(1 - \gamma_\infty). \qquad (42a)$$

Further simplification is evident for γ_∞ may be assumed zero since the very small values of a indicate negligible shielding effects due to adjacent oxygen ions (Abraham et al. 1975). Values of q_P calculated from (42a) have been compared with simple determinations of the field gradient along the defect axis assuming point charges. The measured values of q_P are consistently more negative than the calculated values; this is consistent with a significant contribution to q_P from nearest neighbour O^{2-}, as observed also in the case of the axial ^{25}Mg q_P values for the V^- centre in MgO. To properly account for the distortion will clearly require more knowledge of the displacements of the neighbouring oxygen cores.

One other interesting feature of the hyperfine structure of the trapped hole centres is the relative sign of a and b. For V_{Al}, V_{OH}, V_{OD} and V_F centres these parameters are of the same sign—positive. However all other cases are consistent in that a and b of opposite sign with b the dipolar constant being determined by the sign of g_N. Thus the sign of a is 'anomalous', implying that $|\psi(0)|^2$ is negative. This indicates that there exists a mechanism which transfers more upspin ($\uparrow$) than down spin ($\downarrow$) to the alkali nuclei, since we can write

$$ha = \frac{8\pi}{3} g_N \beta_N g\beta \{|\psi(0)\downarrow|^2 - |\psi(0)\uparrow|^2\}.$$

Schirmer (1968 and 1971) has considered several mechanisms which might explain the *negative* charge density at the Li^+ ion. Neither exchange polarization nor covalent spin transfer of $2s$(Li), $2p$(Li) up-spin ($\uparrow$) are effective enough to determine alone the sign of a. However when indirect overlap with neighbouring O^{2-} ions is included the sign of a is found to be negative for $|Li^0|$ centres and of approximately the correct magnitude (Schirmer 1973). Similar calculations for $|Na^0|$ and $|K^0|$ centres have yet to be attempted.

The g-shifts have already been commented upon briefly. If the evidence of Rose and Cowan (1974) that $A(p_z) \to E(p_x, p_y)$ transitions of V^- centres in MgO occur at a much lower energy than the very strong observed optical band extends to other hole centres good agreement may be obtained between the measured and theoretical $g_\perp$-values. However the values of $g_\parallel$ are worthy of some consideration since as we noted earlier second order spin-orbit interaction gives a shift $\Delta g_\parallel \approx -\frac{1}{4}\Delta g_\perp$. The results show that there is some *positive* shift in the $g_\parallel$-value, which has not yet been fully accounted for, although hole transference among all the oxygen ions will contribute positively to $g_\parallel$ (Tohver et al. 1972).

Clearly there remain several unexplained features, viz. g-shifts, distortion parameters, Fermi contact interaction, oscillator strengths and the nature of the excited states, which must be accounted for in any realistic model of these trapped hole centres. Recent theoretical studies by Schirmer et al. (1974) and Schirmer (1976) offers the most promising prospect of a reasonable explanation of the optical properties of V^- centres. His model treats the V^- centre as a bound small polaron subject to a pseudo Jahn–Teller effect, which stabilizes the hole at one of the six equivalent sites for times long on the ESR time scale. The optical transition is between a ground p_z state on one site to a state comprising a linear combination of p_z functions on two opposite sites in the plane perpendicular to the defect axis: in this sense we are involved with a charge transfer transition. Schirmer shows that the V^- band should comprise two overlapping Gaussian bands, one polarized parallel to the z-axis with peak at $2(E_{JT} + J)$ and the other x, y polarized with peak at $2E_{JT}$, where E_{JT} is the Jahn–Teller stabilization energy and J is the resonance integral. The transition moment is shown to be

$$M = \langle \psi_e | -e\vec{r} | \psi_g \rangle \simeq Je\vec{l}/2E_{JT}$$

where $\vec{l}$ is a vector connecting the two O^{2-} ion sites. Obviously M can be large since $\vec{l}$ is large. Thus a particular difficulty in one's understanding of the V^- centres, viz. the large oscillator strength, is readily understood. However most impressive is the fact that the experimental band peak, shape and strength may be reconciled with theory not just for V^- centres but also for $[M]^0$ centres, V_{OH} centres in MgO, CaO and SrO.

§ 5. THE STRUCTURE AND PROPERTIES OF ELECTRON EXCESS CENTRES

It was noted in § 1.5 that when an alkaline earth oxide is simultaneously subjected to fast-neutron and γ-irradiation in a reactor, one of the products is an F^+ centre. Since it is known that γ-irradiation itself does not lead to F^+ centres in MgO or CaO, it is reasonable to conclude that atomic displacements by neutrons lead to the anion vacancies necessary for F^+-centre formation. It is likely that the atoms are displaced to interstitial sites; heating of reactor-irradiated crystals of MgO has been shown to produce interstitial dislocation loops (Bowen and Clarke 1964).

Since doses in excess of 10^{20} nvt lead to approximately 10^{19} F^+ centres, as monitored by the ESR spectrum, it is assumed that there are approximately this number of (presumably isolated) anion vacancies produced. Usually, X-irradiation of an as-received neutron-irradiated MgO crystal does not greatly increase the number of F^+ centres: hence the number of F^+ centres observed may be taken as an approximate measure of the number of isolated anion vacancies produced. 'Presumably isolated' is to be interpreted to mean only that the trapped electron has six Mg^{2+} neighbours. The ESR spectrum of the F^+ spectrum is insensitive to moderate distortions; the extremely small difference $g_\perp - g_\parallel$ resulting from the association of a cation vacancy with the anion vacancy in which the electron is trapped (P^- centre) testifies to this distortion insensitivity (§ 6). In fact, the S-centre (§ 8), which is presumably an electron trapped in an anion vacancy at the surface of MgO, shows an apparently isotropic g factor of 2·0007. The distortion insensitivity fortunately makes the intensity of the F^+-centre spectrum a fair measure of the isolated-anion-vacancy concentration. The occurrence of optical absorption bands in the visible region even at low doses is interpreted as evidence for the formation of a number of different F-aggregate centres (§ 6). The extent of this aggregation can be minimized by cooling the sample during reactor irradiation.

Unfortunately, one has no counterpart ESR spectrum to assess the number of cation vacancies generated by neutron irradiation. The V^- centre described in § 4 is not seen even with subsequent X-irradiation unless the fast neutron dose is small (of the order of 10^{17}), or subsequent annealing is undertaken. Impurity spectra, even those with isotropic g factor, tend to disappear at doses of 10^{20} nvt. They reappear upon annealing at progressively higher temperatures.

The concomitant radioactivity with even moderate reactor irradiation is an inconvenience. This is especially true with SrO and BaO, for which one must safeguard against long-lived products generated by the slow neutrons. For MgO no satisfactory alternative radiation technique has been reported. However, for CaO, SrO and BaO, the use of a 2·5 MeV proton beam has been effective in generating F^+ centres (Bessent et al. 1968). In all of these, no difference was detected in the ESR spectrum of the F^+ centres generated by neutron or proton irradiation. (It is stated that the other ESR spectra seen in CaO and SrO were different in the proton-irradiated samples; the ESR spectrum of the F^+ centre is the only one seen in BaO).

F^+ centres may also be produced by additive coloration in each of the four oxides: a complication arises in that F centres are also produced. In MgO this led to some confusion about the assignment of the F/F^+ band because the two charge states of the anion vacancy, i.e. the F and F^+ centres, absorb optical photons of almost the same energy. Much the same is true of anion vacancy centres produced in MgO by 2 MeV electron irradiation, where failure to observe the ESR spectrum in crystals with the characteristic 5·0 eV optical band stimulated considerable experimental and theoretical activity. Because of the essential simplicity we will start by reviewing the ESR spectra of F^+ centres, then discuss the optical and magneto-optic properties of both F^+ and F centres.

5.1. *The ESR and ENDOR spectra of F^+ centres*

The ESR spectrum of the F^+ centre in MgO is beautifully simple and unambiguous. Its interpretation requires little more than a table of nuclear spins and of natural abundances. An electron trapped in a negative ion vacancy will have as its neighbours mostly ^{24}Mg and ^{26}Mg, since their combined natural abundance is 89·89%. The nuclide ^{25}Mg constitutes the remaining 10·11% and has a nuclear spin of $\frac{5}{2}$, while for the other nuclides it is zero. Only $(0.8989)^6 = 0.52$ of all trapped electrons will have only ^{24}Mg or ^{26}Mg as neighbours. These centres will give rise to a single line. Of the remaining F^+ centres, the fraction 0·356 will have one ^{25}Mg neighbour; the behaviour of the latter is given by the spin Hamiltonian:

$$\mathcal{H} = g\beta \vec{B} \cdot \vec{S} + A_\perp(S_x I_x + S_y I_y) + A_\parallel S_z I_z. \tag{10a}$$

If B is parallel to one principal axis of the MgO crystal, the hyperfine spectrum will consist of two sextets with couplings $A_\parallel$ and $A_\perp$. The lines corresponding to the latter will have twice the intensity of the former. At a general orientation, three sextets will be present. With $B\|[111]$, the three sextets have equal splittings, as shown in fig. 20. The hyperfine splitting within a sextet is given approximately by $3.94 + 0.48(3\cos^2\theta - 1)$ G, where θ is the angle between the field and the axis containing the ^{25}Mg atom responsible for the splitting. A fraction 0·100 of all the F^+ centres

Fig. 20

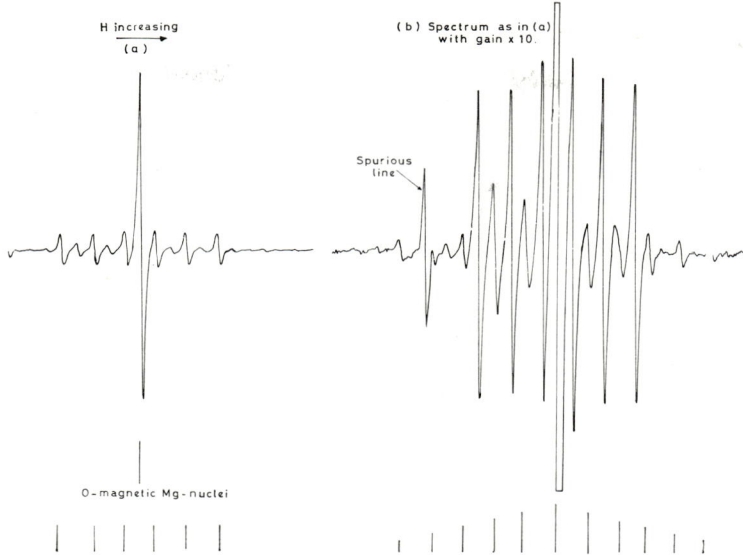

The ESR spectrum of the F^+ centre in magnesium oxide with H_0 parallel to a [111] axis, and showing clearly resolved nuclear hyperfine interaction with (a) one ^{25}Mg nucleus and (b) (at high gain) two ^{25}Mg nuclei.

will have two ^{25}Mg neighbours. In the simplest cases the two nuclei lie on opposite sides of the trapped electrons and give an 11-line hyperfine splitting pattern corresponding to a system with $I = 5$ and having axial symmetry. The splitting will be different for the ^{25}Mg nuclei if they are not co-axial with the trapped electron, unless the magnetic field lies along one of the four body-diagonal directions. Then all hyperfine lines from the two equivalent nuclei will contribute to a single group of eleven lines of relative intensity $1:2:3:4:5:6:5...$. The central line will underlie the single line from the ^{24}Mg and ^{26}Mg nuclides. The remaining ten lines lie midway between or outside the sextet from F^+ centres having one neighbouring ^{25}Mg atom. Some of these are readily discernible in fig. 20. Within the accuracy of measurement, the intensities of all the lines are in accord with predictions based on natural abundances (Wertz et al. 1957). The g factor and the hyperfine splitting parameters are given in table 10 for all of the F^+ centres in the alkaline earth oxides.

Under ordinary conditions, the F^+ centre ESR spectrum in CaO is free of obvious hyperfine splitting because of the low (0.13%) natural abundance of the ^{43}Ca nuclide with $I = \frac{7}{2}$. However, the remarkable aspect of this line in samples prepared under reducing conditions is its very small line width, viz. <20 mG. If a modulation frequency of 100 kHz is used for the sake of low crystal detector noise, the line appears to have structure because of inevitable modulation sidebands. However, if the modulation frequency is reduced to 15 kHz, a line of simple shape and < 20 mG width is observed. Recently, a hyperfine octet has been detected from one ^{43}Ca nearest neighbour in natural abundance. A hyperfine sextet from next-nearest ^{17}O atom, also in natural abundance is discernible. The splitting constants of ^{43}Ca are $A_\perp = 11 \cdot 11$ G and $A_\parallel = 8 \cdot 20$ G, which agree well with data from powder enriched with 2% ^{43}Ca (see table 10 and Tench and Nelson 1967).

The F^+ centre spectrum in SrO corresponds to the same spin Hamiltonian as that for MgO. The even nuclides of strontium (mostly ^{86}Sr and ^{88}Sr) give rise to an intense isotropic central line. Hyperfine splitting arises from ^{87}Sr nuclei with $I = \frac{9}{2}$, with natural abundance 7·14%. For the field along the [111] direction, there are ten lines from one nearest-neighbour ^{87}Sr nucleus. The isotropic splitting parameter $a = -14·65(8)$ G and the dipolar parameter $b = -1·55$ G. Lines occurring between pairs of this set of ten lines might be expected to arise from F^+ centres with two ^{87}Sr nuclei. However, the intensity of these lines is far too great for such an explanation. Instead, these are believed to be $\Delta M = \pm 1$, $\Delta M_I = \pm 1$ transitions. Weaker lines outside the primary group of ten are of the right intensity to arise from two ^{87}Sr nuclei such that ΣM_I for the pair changes by ± 1. The high intensity of the forbidden lines is found to be consistent with the mixing of states produced by the interaction of the large quadrupole moment of Sr with a large electric field gradient in SrO (Culvahouse et al. 1965).

The ESR spectrum of the F^+ centre in BaO shows marked hyperfine contributions from the ^{135}Ba (abundance 6·59%) and ^{137}Ba (11·32%), both of which have $I = \frac{3}{2}$. The central line is the unsplit component from the even nuclides of Ba, and should correspond to 31% of the total F^+ intensity. If one ignores the differences in the nuclear moments (respectively 0·837 and 0·936 nuclear magneton), there should be a hyperfine quartet from those F^+ centres having one magnetic nucleus as nearest nearest neighbour. Forty per cent of the total F^+ intensity should be in this quartet. Twenty-two per cent should be in a septet arising from F^+ centres containing two magnetic nuclei. Separate splittings have actually been determined for the ^{135}Ba and ^{137}Ba nuclei (Mann et al. (1969)). These data are given in table 10.

The spectra reported for MgO, CaO, SrO and BaO show no evidence of effects due to more remote shells of nuclei in contradistinction to the situation in halide lattices (Henderson and Garrison 1973). However as can be seen in table 9 the F^+ centre spectra are characterized by g-values which are less than 2·00 and hyperfine parameters which increase rapidly with isotopic mass.

As we have seen in Chapters 2 and 4 the hyperfine terms give direct information about the radial extent of the wave function and about the distortion of the lattice in the neighbourhood of the defect. Many calculations on F centre properties use relatively simple envelope functions (see for example Gourary and Adrian 1957): they give quite good values for the total energy of the electronic states. Inevitably the early calculations neglected the rapid fluctuations of the F^+ centre wave function on the neighbouring ion cores. Accurate calculations of the hyperfine parameters

Table 10. First shell hyperfine structure data for F^+ centres in alkaline earth oxides.

Host	g-value	a/h (MHz)	b/h (MHz)	A_R	$\|\psi_F(0)\|^2 \times 10^{-21}$ (cm^{-3})	Reference
MgO	2·0023	−11·03	−1·33	325	0·84	Unruh and Culvahouse (1967)
CaO	2·0001	−25·06	−2·71	700	0·83	Henderson and Tomlinson (1969)
SrO	1·9845	−40·775	−4·279	1850	0·77	Culvahouse et al. (1965)
^{135}BaO	1·9355	184·7	18·4	3300	0·88	Mann et al. (1969)
^{137}BaO	1·9355			3300	0·88	Mann et al. (1969)

are only possible if such core oscillations are present in the wave function (Gourary and Adrian 1957; Halliburton *et al.* 1976; Wood 1970; Kemp and Neeley 1963a; Allsop *et al.* 1973). This can be done by orthogonalizing a smooth vacancy centred envelope wave function, ψ_F, to $\{\psi_i\}$ the set of near neighbour ion core orbitals as in eqn. (16). As noted in Chapter 2 this has the effect of modifying the Fermi contact in such a way that $|\psi(0)|^2$ is replaced by $A(R)|\psi_F(R)|^2$. The first shell data are summarized in table 9, the values of A_R being extrapolated from alkali halide values (Gourary and Adrian 1960) using the $(Z-2)^{3/2}$ relationship (Hughes and Henderson 1972, Henderson and Garrison 1973). It is notable that the values of $|\psi_F(R)|^2$ are closely similar but much smaller than their counterpart in the alkali halides. This is to be expected since the excess charge on the F^+ centre concentrates the wave function in the vacancy to a greater extent that is apparent in the alkali halide F centres.

The ESR spectrum leaves little doubt about the identification of the F^+ centres and the extent of wave function ψ_F overlap on the nearest neighbour nuclei. However the ENDOR spectrum is equally important because it may give information about the other shells of nuclei and, through quadrupole terms, the displacements of neighbouring ions. In fact in the ESR spectra of F^+ centres in SrO and BaO one necessarily must include the effects of quadrupole interactions. For SrO the value of P was computed from measurements of the positions and intensities of the forbidden lines (Culvahouse *et al.* 1965). Such forbidden lines are not observed in BaO: Mann *et al.* (1969) determined values of P for both ^{135}Ba and ^{137}Ba by computer diagonalization of the secular equation at orientations where least scatter on the experimental results was obtained. Nevertheless more accurate values of all spin Hamiltonian parameters are obtained from the ENDOR measurements. So far first shell ENDOR spectra have been reported for MgO and SrO. The measured value of P in MgO coupled with the quadrupole moment of the ^{25}Mg nucleus with a value of $Q = +0.22 \times 10^{-24}$ cm^2, leads to an electric field gradient at the nearest magnesium atoms of:

$$q_1 = 3.40(2/R_1^3),$$

where R_1 is the Mg-O distance, viz. 0.21 nm. The use of an electric field gradient expression proposed by Feuchtwang (1962) and modified by Culvahouse *et al.* (1965) leads to that part q_1 (pol) of the electric field gradient due to lattice distortion. The value is given as:

$$q_1(\text{pol}) = -0.238(2/R_1^3).$$

This large negative value is interpreted as indicating a relatively large outward displacement of the Mg^{2+} ions which are nearest neighbours of the trapped electron. The result is in accord with theoretical computations of Kemp and Neeley (1963 a). The ESR experiment with F^+ centres in SrO (Culvahouse *et al.* 1965) leads to a nearly identical lattice-distortion-produced electric field gradient:

$$q_1(\text{pol}) = -0.240(2/R_1^3).$$

A similar result obtains for q_1(pol) in BaO (Mann *et al.* 1969). Taken together the quadrupole interactions imply a lattice relaxation of 5–8% outwards for first shell neighbours in these three oxides.

Recently there have been reports of ENDOR results on the third and fifth shells of ions (Mg^{2+}) by Halliburton *et al.* (1976) and second, sixth and eighth shells (O^{2-}) by Allsop *et al.* (1973), the locations of which are shown in fig. 21 (*a*). Crystals enriched

Fig. 21

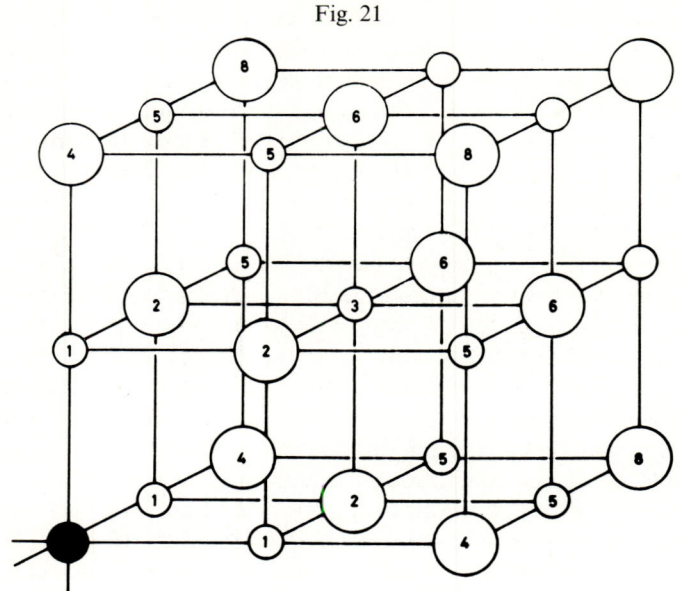

(a) Showing the positions of various shells of ions neighbouring the F^+ centre in MgO.

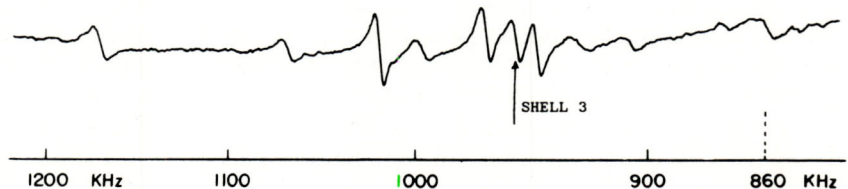

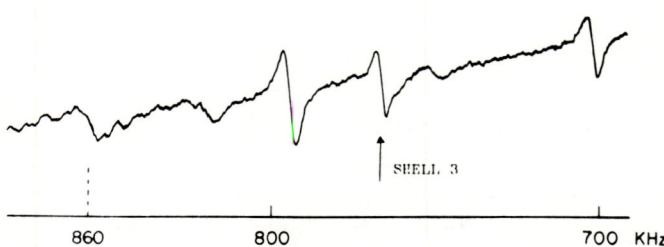

(b) The ENDOR spectrum of F^+ centres in MgO showing the lines due to ^{25}Mg in shell 3 and some lines due to shell 5 (after Halliburton et al., 1976).

with ^{17}O (Hann et al. 1971) were used in this latter case and values were obtained for the isotropic contact term and the radial part of the anisotropic constant. No quadrupole splittings were observed for the ^{17}O spectra, thus there can be no direct estimate of the O^{2-} ion displacement. The ENDOR spectrum on the ^{25}Mg nuclei do give quadrupole couplings in addition to the isotropic and anisotropic hyperfine coupling terms. Some of the ENDOR lines for the third and fifth shells are shown in fig. 21 (b) for an orientation in the (100) plane with along the [100] direction. Orientation dependence studies show that the two lines at 763 kHz and 956 kHz belong to the third shell: all other lines in this spectrum are from the fifth shell ions. The twenty-four nuclei in this shell are divided into three groups of eight nuclei in this orientation: within each group the ENDOR spectra are equivalent giving rise to a total of thirty lines approximately symmetric about 860 MHz, the free ^{25}Mg nuclear resonance frequency. Nineteen of these lines have been identified by Halliburton et al. (1976) who also determined the relative signs of the spin Hamiltonian parameters by selective saturation of different parts of ESR line (Spaeth 1966). The remaining lines are overlapped by much more intense lines observed above 1370 KHz.

Halliburton et al. fit these spectra to a spin-Hamiltonian modified slightly from eqn. (11):

$$\mathcal{H} = A\vec{I}.\vec{S} + \vec{I}.\tilde{B}.\vec{S} + \vec{I}.\tilde{Q}.\vec{I} - g_N\mu_N\vec{B}.\vec{I}.$$

The traceless tensors $\tilde{B}$ and $\tilde{Q}$ are symmetric requiring five independent parameters for their description in their respective principal axis system. In the first and third shells symmetry considerations reduce the number of parameters required to a, b and P. The only symmetry element for a fifth shell site is a reflection plane: a principal axis of $\tilde{B}$ and $\tilde{Q}$ must be perpendicular to this plane. Angles α_B and α_Q fix the orientation of the other principal axes of $\tilde{B}$ and $\tilde{Q}$ in the mirror plane. The data for the fifth shell then give the parameters a, b and b' to describe the hyperfine splittings and the quadrupole components P and P'. Note that the primed parameters display the essential departures from axial symmetry in the appropriate principal axis system. The best fit values for these spin-Hamiltonian parameters are shown in table 11 together with similar data for the ^{17}O nuclides in the second, sixth and eighth shells.

The physical significance of the quadrupole term lies in the magnitude of the induced electric field gradients at the particular sites. Again the analysis by Feuchtwang (1962) is adapted to the divalent lattice to obtain values for $q(\text{pol})$, that part of the field gradient due to the distortion of each shell of ions (Halliburton et al. 1976). The magnitude of $q(\text{pol})$ can then be related to the ionic displacements and the electronic polarizations of the various shells using the methods of Das and Dick (1962) and Dick (1966). From the third shell data the displacements ξ_i of shells one to five

Table 11. Comparison of Experimental and Calculated Hyperfine Parameters†

Shell	a (MHz) Expt.	a (MHz) Theory	b (MHz) Expt.	b (MHz) Theory
1 (Mg)	−11·03	−12·4 (b)	−1·33	−0·698 (a)
2 (O)	−13·9	−12·4 (b)	−2·85	—
3 (Mg)	−0·196	−0·463 (a)	−0·075	−0·076
5 Mg	−0·218	−0·238 (a)	−0·056	−0·049 (a)
6 O	−0·283	−0·287 (b)	−0·123	—
8 O	−0·540	−0·541 (b)	−0·134	—

(a) Calculated using the envelope function $\psi_F = N(1 + \gamma r)e^{-\gamma r}$ orthogonalized by the Schmidt process (Eq. 16) to the ion core orbitals on Mg^{2+} and O^{2-} (Halliburton et al. 1976).

(b) Calculated using the L.C.A.O. wavefunction

$$\psi_F = \alpha \sum_1^{12} 2p(O^{2-}) - \beta \sum_1^{6} 3s(Mg^{2+}) + \gamma \sum_1^{6} 3p(Mg^{2+})$$

(after Allsop et al. (1973)).

are obtained as the only solutions of physical significance for each sign of P. The first shell moves outward by $\sim 6\%$ and there are smaller outward displacements of shells three, four and five. The second shell displacements are less clearly defined since there is no determination of the electronic polarizability of the second shell oxygen ions. It is estimated that these results imply a nett volume of formation of 4 cm^3/mole. The impressive feature of these results is not simply that the distortion of the first five shells of ions is obtained from the value of P_3 but that they are obtained independent of similar data from ^{17}O hyperfine structure.

There are two important factors which emerge from the hyperfine interaction constants in table 11. The measurements on ^{17}O nuclides by Allsop et al. (1973) demonstrate quite forcibly that a significant overlap of the F^+ centre wave function onto second shell O^{2-} ion core orbitals must be recognized in any meaningful theory. There is also a relatively large isotropic interaction with the fifth shell nuclei which also arises out of indirect overlaps involving the second neighbour oxygens. This is shown in table 11 where we compare the measured hyperfine parameters for shells 1 to 8 with calculated values using a Schmidt orthogonalized wave function (Halliburton et al. 1976) and an L.C.A.O. function (Allsop et al. 1973). The agreement is especially good in the shells 5, 6 and 8 which are so sensitive to indirect overlap through the second neighbour O^{2-} ion.

5.2. Optical properties of the F^+ centres

The alkaline earth oxides are very strongly coloured after even moderate doses of fast neutrons. Dose-dependence studies of the coloration show that the primary effects are due to absorption by intrinsic lattice defects, although changes in the valence states of various impurities may also be important. Magnesium oxide and calcium oxide have been studied in greatest detail but recent studies have also been made of strontium and barium oxides. Typical spectra for these oxides over the wave length range 1800–10 000 Å, reveal that the most prominent optical absorption bands occur at 4·90 eV in MgO, 3·65 eV in CaO, 3·0 eV in SrO and 2·0 eV in BaO. These bands are believed to be due to F^+ centres. It is notable that bands occur at these positions in additively coloured crystals, supporting the suggestion in §1.5 that the primary products of additive coloration are anion vacancies. Apart from the asymmetric F^+ band in barium oxide there is little else in the spectrum to indicate the presence of lattice defects. This is also the case for strontium oxide, at least in the dose range discussed here. It would appear for these two solids that the optical transitions of other defects are hidden by processes concerned with the conduction band and the reststrahlen absorption. For magnesium oxide and calcium oxide numerous other absorption bands are observed.

Historically, the optical absorption bands of the F centres in the alkali halides were rather well studied long before the deBoer model had been fully accepted or before there were ESR and ENDOR techniques available to corroborate the model. However, for the F^+ centres in the alkaline earth oxides, the order of study was inverted. For all the oxides, as well as for the analogous sulphides, selenides and tellurides, the ESR spectrum was first identified. It was some years later that the absorption bands were identified, and this came after some guidance was available from theory (Kemp and Neeley 1963 a, Kemp 1963). Since the optical characteristics are rather different for the F^+ bands in the several alkaline earth oxides, they will be considered separately.

5.2.1. Magnesium oxide

Although there is an extensive literature of radiation coloration in magnesium oxide, no correlations of absorption bands with specific defects were achieved until the last several years. This was in part due to the confusing effects of impurities, especially Fe^{3+}, referred to in § 3. However, it is now known (Soshea et al. 1958) that Fe^{3+} absorbs at photon energies peaking at 4·3 eV and 5·7 eV. In addition to these bands, absorption bands at 4·95 eV, 3·6 eV, 2·15 eV and 1·2 eV are observed in neutron irradiated magnesium oxide. These bands are apparently due to intrinsic lattice defects. They have a decreased intensity for a particular neutron dose if the crystals have been previously heated at about 1200°C in hydrogen. Some interesting phenomena observed in the optical absorption spectrum of heavily irradiated single crystals have been reported by Henderson and King (1966) and by King (1967). The change in the absorption spectrum in the region of the F^+ band is shown in fig. 22 for the dose range 10^{15}–10^{20}. It is apparent that the Fe^{3+} absorption peaks at 4·3 eV and 5·7 eV have reached maximum intensity at a dose level of about 7×10^{16} nvt > 1 MeV. As the dose is further increased the F^+ band grows very rapidly, this growth being accompanied by further structure at higher doses. In fig. 22 it is evident that the F^+ band broadens and that new bands appear at about 4·5 eV and 5·6 eV in crystals irradiated to higher doses. The progressive increase in absorption above 6·0 eV also indicates the presence of absorption bands in the ultra-

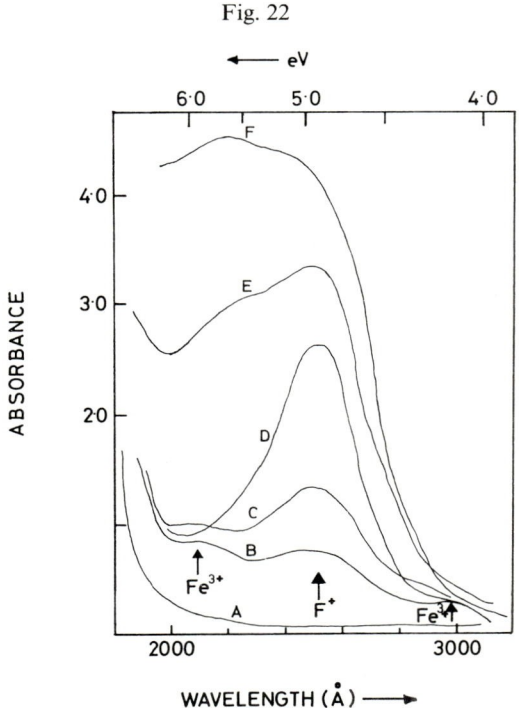

Fig. 22

Changes in the optical absorption spectrum of magnesium oxide in the region of the F^+ band, with increasing fast neutron dose. Sample thicknesses (cm), A: 0·013, B: 0·03, C: 0·012, D: 0·013, E: 0·0025, F: 0·003. Dose (nvt), A: unirradiated, B: 7×10^{15}, C: 7×10^{16}, D: $3·22 \times 10^{17}$, E: $6·8 \times 10^{19}$, F: $1·18 \times 10^{20}$. (after King 1967.)

violet spectrum. Henderson and King (1966) suggested that these new features are due to the formation of vacancy aggregate centres at the higher doses. It is at these high doses that the paramagnetic aggregate defects were first identified.

In the visible region of the spectrum the coloration is very sensitive to the temperature of irradiation. This implies that aggregation results from thermal migration and radiation annealing rather than by the overlap of different displacement sequences. At 45°C (Henderson and Bowen, 1971) only the bands at 3·6 eV, 2·15 eV and 1·2 eV are observed and the growth of these bands is not accompanied by the emergence of other bands. This is contrary to the behaviour observed in fig. 23 for crystals irradiated at 150°C, in which the band at 2·15 eV is observed to be overlapped by other bands which increase in intensity as the dose increases. King (1967) speculated that these bands were due to the aggregation of defects as the dose increases. However, at no point do these bands become clearly resolved and it may well be that the broad structure in the region of the 2·15 eV band is associated with close neighbour defects which shift the band peak as a result of a lowering of the symmetry around the defect. The bands at 3·6 eV and 1·2 eV change intensity in parallel fashion as a function of dose, and are apparently due to two different transitions within the same centre (§ 6).

Fig. 23

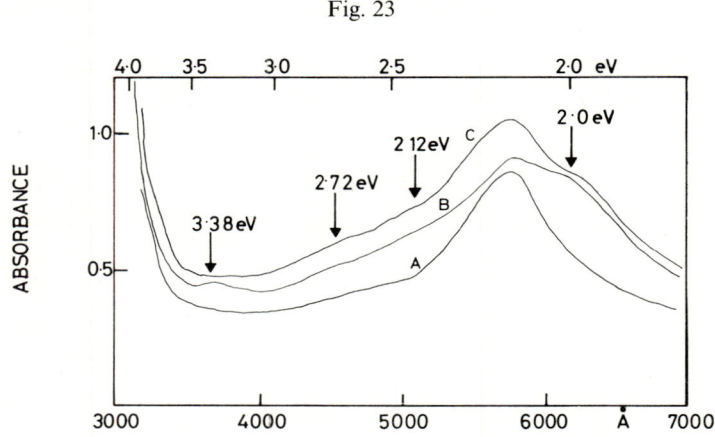

The visible region of the optical absorption spectrum of neutron-irradiated magnesium oxide for different fast neutron doses. The broad peaks indicated are observed only in crystals irradiated to doses such that the F^+ centre has attained a saturation concentration. These bands are close to the positions of aggregate bands in annealed crystals. Doses (nvt), A: $6·8 \times 10^{19}$, B: $1·18 \times 10^{20}$, C: 2×10^{20}. (After King 1967.)

At even higher temperatures, 400–600°C, the coloration is much less than for the same dose at low temperatures. In addition, the absorption spectrum is more typical of crystals which have been annealed after irradiation at temperatures in the range 400–700°C. It is quite evident that at these temperatures radiation annealing very strongly reduces the radiation damage. This is due partially to recombination processes, as well as to the growth of dislocation loops which are formed during irradiation. In addition the vacancies aggregate to such an extent that only about 10^{15}–10^{16} F^+ centres are observed after irradiating at 600°C to a dose of $4·3 \times 10^{20}$ nvt.

At this point it is appropriate to comment on the extent of the damage process as measured by colour centre techniques. It is clear that the predominant defect is the F^+ centre. The concentration of F^+ centres after a dose of 10^{19} nvt as measured either by paramagnetic resonance or by optical absorption is about 10^{19} F^+ centres/cm^3. Presumably, there are also other anion defects but after ascribing all the absorption to such defects it is apparent that the damage on the anion sublattice cannot represent much more than about $1\cdot5$–$2\cdot0 \times 10^{19}$ vacancies/cm^3. The measured damage is thus less than 1% of the damage calculated on a simple Kinchin and Pease model (1955). This surely indicates the need to study the damage process at lower temperatures to preclude the possibility of interstitial migration and of radiation annealing.

A comparison of the intensity of the MgO F^+-centre ESR spectrum with that of any of the optical bands in the visible or near-ultra-violet region clearly shows that these bands cannot represent F^+-centre absorption. However, gaussian analysis of the absorption in the 4–6 eV region shows a peak near 5·0 eV which appears to decay upon thermal bleaching in parallel with the decay of the ESR signal of the F^+ centre (Wertz et al. 1964). This is remarkably close to the value of 4·7 eV predicted by Kemp and Neeley (1963 a). The first record of observation of what later turned out to be the F^+ optical band in MgO is found in unpublished work of Boyd et al. (1947), and also of Nelson and Pringsheim (1948). The former found that a band near 5 eV was required to explain their absorption spectra after neutron irradiation. The latter found a partially resolved band at 5 eV for MgO crystals which had received a moderate neutron dose. No attempt was made at interpreting this band. Clarke (1957) reported this band along with numerous others in neutron-irradiated MgO.

A detailed comparison of intensities of the 4·95 eV band and the number of F^+ centres determined by the ESR absorption has been carried out for samples subjected to a variety of neutron-irradiation treatments (Henderson and King 1966, Henderson et al. 1968). The intensity of the peak at 4·95 eV has been obtained by gaussian analysis, allowing for peaks at 4·3 and 5·62 eV for Fe^{3+} (Soshea et al. 1958, Peria 1958), as well as for lines of fixed half-width at 5·30, 5·58, 5·95 and 6·8 eV (Johnson 1954). The half-widths assigned to these peaks at 4°K were 0·23 eV, 0·305 eV and 0·73 eV. The half-widths of the Fe^{3+} lines are respectively found to be 0·73 eV and 0·96 eV. It has been shown that for doses less than 10^{18} nvt (counting neutrons having energies in excess of 1 MeV) the position and half-width of the 4·95 band are well defined. The most recent value of the half-width of this band, obtained by least-squares computer analysis, is 0·477 eV at 4°K. One may use Smakula's formula relating the number n_F of F^+ centres and the absorption coefficient α_F at the peak maximum to determine the oscillator strength f:

$$n_F f = 0\cdot 87 \times 10^{17} \frac{n}{(n^2 + 2)^2} \alpha_F W(T). \qquad (20)$$

Here the index of refraction, n, is taken as 1·74; the oscillator strength f is given as 0·80 if the half-width $W(T)$ is taken as 0·67 eV at $\sim 290°$K (Henderson and King 1966).

At one time the assignment of the 4·95 eV band as the F^+ band appeared controversial. MgO samples additively coloured in Mg vapour or subjected to fast electron irradiation were found to show this band strongly. However they did not show an F^+ ESR spectrum (Chen et al. 1968 a). The theoretical predictions of Wood and his associates (Wood and Opik, unpublished, Wood and Wilson 1975) were crucial in satisfactorily resolving the problem. These calculations predict that absorption bands due to F and F^+ centres should occur at almost the same photon energy. The conclusion drawn from these studies is that neutron irradiation produces largely F^+ centres whereas additive colouration or electron irradiation gives rise to large concentrations of F centres.

Measurements of the temperature dependence of the F^+ band may be used to obtain insight into the vibronic coupling in these centres (see § 2.2.1). From the half-width data, using equations relating the second moment $M_2(T)$ to the absolute temperature, one may obtain values for the Huang-Rhys factor, S, and the effective frequency, ω, of phonons coupled to the electronic states. Henderson et al. (1968) find $S = 39$ and $\hbar\omega = 260\,\text{cm}^{-1}$: however as noted above to do so they analysed the absorption spectrum in the neighbourhood of the F^+ band in terms of a sum of gaussian bands. Subsequent work by Kappers et al. (1970) and Chen et al. (1969a) shows that the bandwidth measured in this way is an underestimate. In rather purer crystals the band is clearly asymmetric in shape: the most reasonable value of the half-width is then 0.57 eV (Chen et al. 1969). The major conclusions, however, are unchanged: the Huang-Rhys factor is large (~ 30), so that the electron phonon coupling is strong. Furthermore the effective phonon frequency tends to lie in the acoustic branch of the phonon spectrum (Peckham 1967).

5.2.2. Calcium oxide

The absorption spectrum of neutron-irradiated CaO shows several strong bands. Of these, a band at 3.7 eV is found also in additively coloured CaO and it likewise is correlated with the F^+-centre ESR intensity (Tanimoto, quoted in Kemp et al 1968). This band also shows reversible bleaching and growth in an $F^+ \leftrightarrow F$-centre transformation, akin to the $F \leftrightarrow F'$ conversions in the alkali halides (Kemp et al. 1967). The position of this band was predicted with remarkable accuracy by use of the same techniques which gave good results in MgO (Kemp 1963). The band is asymmetric in shape: this asymmetry is not changed by annealing (Bessent et al. 1968). The most surprising aspect of this band is the appearance of a zero-phonon line at 355.74 nm which has been assigned as the F^+-centre zero-phonon transition by Kemp et al. (1968). In the alkali halides and MgO zero-phonon peaks are seen only for F-aggregate defects: in these defects the electron–phonon coupling is usually weak. The criterion for weak coupling and therefore for detecting the 0-phonon lines is that the Huang–Rhys factor $S < 6$(§ 2). S can be estimated in a number of ways including the Stokes shift between the band peaks in emission and absorption, the band width at $0°K$, and the integrated zero-phonon line intensity relative to the broad band. Such techniques in this case yield S values in the range 7 to 16 (Henderson and Tomlinson 1969, Bessent, to be published). First observation of emission from the F^+ centre in calcium oxide was reported by Henderson (1967).

5.2.3. Strontium and barium oxides

The spectrum of additively coloured SrO has been reported, but up to 1968 no actual spectra had been published. A symmetrical peak is reported at 2.50 eV with half-width 0.84 eV. This peak was attributed to the neutral diamagnetic centre. Its shift to 3.1 eV on annealing *in vacuo* is presumed to be due to colloid formation (Johnson and Hensley 1967). An unsymmetrical peak at 3.0 eV has been reported for neutron-irradiated SrO. The asymmetry is said to remain as the 3.0 eV band is thermally annealed. Its assignment as the F^+ band in SrO is made on the basis of Faraday rotation studies (Bessent et al. 1968).

The optical absorption of additively coloured BaO was studied before any F^+ optical absorption band had been identified in an oxide. Bands are reported at 1·4 and 2·0 eV (Dash 1953); Kane (1951) places the band at 2·4 eV. It is stated that photoconductivity associated with the 2·0 eV peak is greatly enhanced by X-irradiation (Dash 1953). The band at 2·0 eV was first attributed to the neutral F centre (Sproull et al. 1953). The first ESR experiments were made on BaO heated in Ba vapour or in liquid Ba; these failed to find a correlation of the intensity of the F^+ ESR spectrum with that of the 2·0 eV band. However, the F^+ centres were confined to a region near the surface (Carson et al. 1959). (The analogous lack of correlation of the 5 eV band seen in additively coloured MgO with F^+-centre concentration was mentioned earlier.) However, the Faraday rotation experiments to be discussed in § 6.4 identify the 2·0 eV band induced by 2·5 MeV proton irradiation as arising from the F^+ centre. As noted in § 5.2, no other ESR spectrum is seen in the BaO at 4, 77 or 300°K. In § 4 we observed that the 2·0 eV band obtained by proton irradiation has a long tail at high energies. This is attributed to the close proximity of the excited level to the conduction band. As with the absorption band in SrO and CaO, the shape of the 2 eV band does not change on annealing. Annealing causes parallel changes in absorption intensity as it decays with the ESR absorption (Bessent et al. 1968).

5.3. Magneto-optic and vibronic properties

To a first approximation, one may regard an F^+ centre as being akin to an atom with an S ground state and a P excited state, i.e. an alkali atom. The latter shows a wave length-dependent Faraday rotation of the plane of polarization of a light beam directed parallel to a magnetic field. The rotation occurs because of the difference in the index of refraction of the two components of a plane-polarized beam. The magnitude of the angle of rotation, θ, is given by:

$$\theta = \frac{2\pi l \delta n}{\lambda_0} \approx 2\pi r \frac{dn}{dE} |E^+ - E^-|. \qquad (44)$$

Here l is the absorption path length, δn is the difference in index of refraction for the two circularly polarized components of the incident beam, λ_0 is the wavelength in vacuo of the radiation used, and $E^+ - E^-$ is the difference in Zeeman splitting for the two components of the circularly polarized light. In the presence of a magnetic field, the ground state is a spin doublet; the first excited state is a quartet, and above the latter is a doublet (this applies to the F centre; the opposite is true of an alkali atom). The centres of gravity of the doublet and the quartet are separated by an energy ΔE. (This is an experimental observation which has received considerable theoretical support (Smith 1965).)

Dexter (1958) calculated the Faraday rotation to be expected for F centres in alkali halides at concentrations such that optical absorption experiments can be performed. For a concentration of 10^{18} cm^{-3} he predicted a rotation of -2.55×10^{-3} G^{-1} cm^{-1} at the centre of the absorption band. The predicted shape of the plot of rotation versus energy is that of the derivative of a dispersion curve. Subsequent experiments on F centres in the alkali halides have detected the rotation and have verified the predicted shape of the rotation curve (Mort et al. 1965). However, a marked temperature-independent diamagnetic rotation was also observed. In contrast, the paramagnetic contribution has a $1/T$ dependence, and hence may greatly exceed the diamagnetic contribution for low temperatures. The magnitudes of both paramagnetic and diamagnetic components can be calculated reasonably well from the alkali atom model. It is necessary to take the spin–orbit coupling to be negative in the excited state as described above; this implies that the $^2P_{1/2}$ levels lie above the $^2P_{3/2}$ levels, as mentioned earlier. The value of the rotation at the minimum of the rotation curve is given as:

$$\theta_{\min} = -\left(\frac{\ln 2}{\pi}\right)^{1/2} \frac{\alpha_{\max} l \mu_B B}{W(T)} \left(\frac{g_2 + 5g_3}{3} - \frac{g_1 \Delta E}{3kT}\right). \tag{45}$$

Here $\alpha_{\max}$ is the absorption constant at the centre of the absorption band and $W(T)$ is its full width at half maximum, μ_B is the Bohr magneton, g_2 and g_3 are respectively the Landé factors for the $^2P_{1/2}$ and $^2P_{3/2}$ states, g_1 is the spectroscopic splitting factor for the ground state and ΔE is the separation of the centres of the excited states. More elaborate theories which take into account the effect of lattice vibrations do not markedly alter the magnitude of the predicted rotation (Henry et al. 1965). The first of the two terms in the latter parentheses of the expression for $\theta_{\min}$ represents the diamagnetic contribution, while the latter is the paramagnetic contribution to the total rotation. One of the important benefits of the observation of Faraday rotation is that it permits some description of the unrelaxed excited states, while observation of emission gives information only about the state reached after relaxation.

The observation of Faraday rotation of F centres in the alkali halides has led to studies in MgO, then CaO (Kemp et al. 1966, 1967 a) and finally SrO and BaO (Bessent et al. 1968). The primary aim was to verify that the bands at 4·95, 3·7, 3·0 and 2·0 eV which had been assigned to the F^+ centres in each of these oxides is indeed the F^+ band. A Faraday rotation is unquestionably observed in the appropriate region for each of the oxides. The rotation curve for MgO and CaO is said to have a dispersion shape (Kemp et al. 1967 a); however, for lower optical densities of samples, the shape is dispersion-derivative-like (Bessent et al. 1968). Since there is a contribution to the Faraday rotation from inter-band transitions, and from other absorption bands, it comes necessary to extract the contribution of the F^+ band. One of the significant contributions by Kemp has been the combination of ESR and Faraday rotation experiments. In his experimental arrangements, an oxide sample is mounted in a microwave cavity provided with windows such that the optical absorption and Faraday rotation measurements may be made without removal of the sample. The cavity is then placed at the centre of a superconducting magnet such that the sample may be illuminated with polarized light directed along the magnetic field. The net Faraday rotation is governed by the difference in the numbers of absorption transitions made with $\Delta m_J = +1$ and $\Delta m_J = -1$. Hence, these numbers can be controlled by effecting transitions between the $\pm\frac{1}{2}$ states of the ground level of the F^+ centre. Specifically, the application of sufficient microwave power to saturate the ESR transition will cause the paramagnetic Faraday rotation contribution of the F^+ centre to vanish. Hence if the value of the applied magnetic field corresponds to the resonant value for the microwave frequency used, the F^+ rotation contribution is obtained by taking the difference between the rotation versus wave length curves in the absence and in the presence of a saturating microwave magnetic field.

In the case of the F^+ centre in MgO, the very intense optical absorption beyond 4 eV makes the experiment difficult; however, such difficulties have now been overcome (Kemp et al. 1969) with the result that there is no longer any doubt that F^+ centres absorb at a photon energy of 4.95 ± 0.03 eV.

The Faraday rotation of the F^+ centre in CaO is far more amenable to experimental observation for a number of reasons. Since the F^+ band occurs at much lower energy than in MgO there is a much smaller inter-band contribution to the Faraday rotation. Secondly, there is far less over-lapping by other optical bands of the F^+ band in CaO. Finally, in CaO there is the remarkable advantage of having an exceedingly narrow zero-phonon band at the low-energy side of the F^+ band, viz. at 355.7 nm. This affords the opportunity to perform a Faraday rotation experiment both on the main F^+ band and on the zero-phonon line. A superposition of two separate rotation experiments is given in fig. 24 which also shows the absorption spectrum. Here the paramagnetic contribution is obtained for the broad component by taking the difference between the rotation in the presence and in the absence of a saturating microwave field at each of a series of wave lengths of the polarized light. For the zero-phonon line, the wave length was scanned continuously. Bessent et al. (1968) found that for lightly irradiated (10^{17} nvt) CaO and SrO crystals the Faraday rotation curves has a dispersion-derivative-like shape characteristic of the alkali halide F centres. However, no Faraday measurements were attempted on the zero phonon line. The same shape is observed for the 2.5 MeV proton irradiated BaO.

Fig. 24

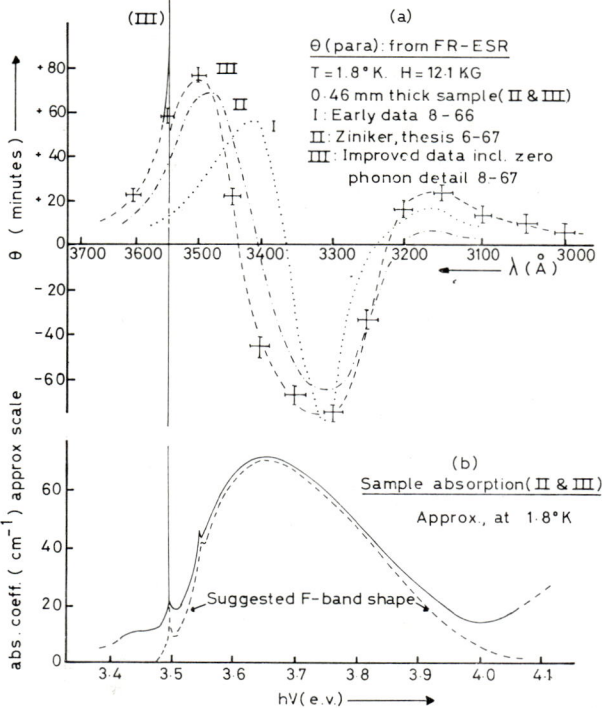

(a) Paramagnetic Faraday rotation of the F^+ band in CaO and of the 3557.4 Å zero-phonon line. Curves I, II and III were obtained by observing the rotation changes at discrete wavelengths λ. The rotation around the zero-phonon line was determined by a high resolution continuous scan and superimposed on III. (b) shows the optical absorption spectrum of the sample. (After Kemp et al. 1968.)

In this case the Faraday rotation becomes increasingly negative upon approaching the absorption edge, in contrast to other ionic crystals.

Values of the spin-orbit splitting of the excited p levels have been calculated from the magnitude of the Faraday rotation. These values are given in table 12. In each case these are negative, as has been observed for F centres in the alkali halides (e.g. Mort et al. 1965) and as has been predicted (Smith 1965). The wave functions used to compute the λ' values in table 12 were probably too compact.

Table 12. Experimental and calculated values of λ' and absorption band peak of F^+ centre for oxide (after Hughes and Henderson 1972).

	λ' (cm^{-1})		F^+ absorption peak		Refs. to λ'
	Exp.	Calc.	Exp.	Calc.	
MgO	−12	—	4.95	4.7	Kemp et al. (1968)
CaO	−60	—	3.7	3.8	Bessent et al. (1968)
	−24	−20			Bessent et al. (1968)
SrO	−185	−80	3.0	3.4	Bessent et al. (1968)
BaO	−265	−180	2.0	3.0	Bessent et al. (1968)

The Faraday rotation studies on MgO (Kemp et al. 1969) demonstrated one further feature, that the vibronic coupling is predominantly to the non-cubic modes of vibration: this aspect assumes a much greater significance in CaO. As noted already Faraday rotation studies on the F^+ centre broad band in CaO yield values of λ' in the range $-(24-60)$ cm^{-1}, whereas measurements on the narrow line imply $\lambda' \simeq 1$ cm^{-1}. Thus there is a reduction in the spin-orbit coupling parameter by a factor of order 0·05. According to Ham (1965, 1968) such quenching of orbital electronic operators originates in the dynamical Jahn–Teller effect. Since measurements on the broad band reveal the true magnitude of the coupling coefficients (Henry et al. 1965) a comparison of the broad band and zero-phonon line properties under external perturbation will reveal directly the quenching or reduction factors. Unfortunately the experiments of Kemp et al. (1968) lacked high resolution so that no accurate value of the reduction factor, $K(T_1)$, was obtained. As described below subsequent work gives a very clear understanding of the reduction of physical operators by the Jahn–Teller effect. A very detailed discussion of the results and theory is given in Hughes and Henderson (1972).

The important result of the work by Kemp and his colleagues was the assignment of both zero-phonon line and broad band to the F^+ centre. However they failed to observe the rather peculiar, squarer than gaussian, band shape. This F^+ band shape shown in fig. 25, is central to the understanding of the electron-phonon coupling in the F^+ centre in CaO. Generally the band shape is controlled by the strength of coupling to modes of various symmetries. The electronic state involved is described by the irreducible representation T_{1u} which couples linearly to vibrational modes of A_{1g}, E_g and T_{2g} symmetry, each coupling being represented by the Huang–Rhys factors S_A, S_E and S_T. The reduction factors for operators transforming as E_g and T_{2g}, i.e. $K(T_2)$ and $K(E)$, may be determined by comparative measurements of the zero-phonon line and broad band under uniaxial stress. Hughes and Runciman (1969) have studied the response of the zero-phonon line to an externally applied stress. The splittings are observed to be linear in stress so that the stress and strain coupling

Fig. 25

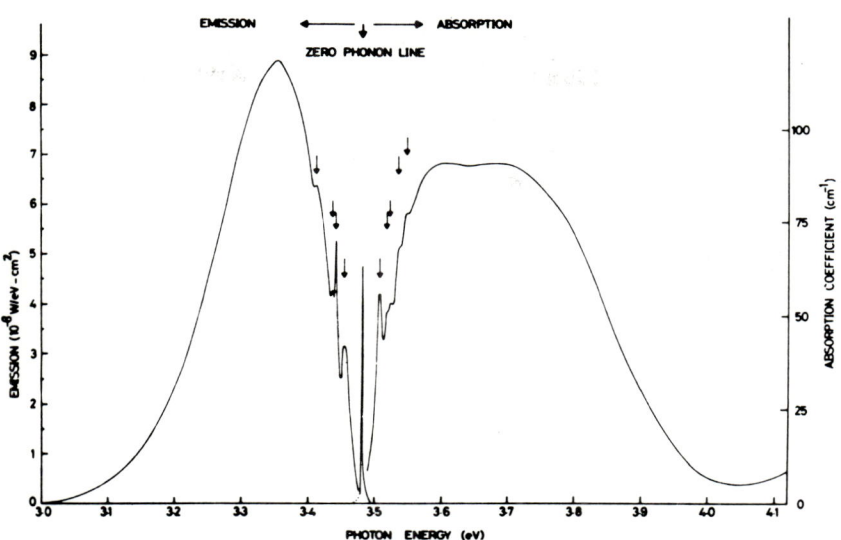

Vibronic structure in the F^+ centre emission and absorption bands in CaO, measured at 4°K (after Henderson et al. 1972).

coefficients are readily deduced (table 13). These coefficients are similar in magnitude to those measured from broad band experiments on F centres in alkali halides. It appears then that these parameters are not much reduced relative to the broad band, in apparent conflict with the Faraday rotation measurements. Furthermore overall values of S determined from the band half-width ($S \approx 13$–16) compare badly with those measured from the fractional intensity in the zero-phonon line or zero-phonon line-band centroid separation ($S \approx 4$–6).

The theoretical work of O'Brien (1969, 1971) provides the physical basis for reconciliation of all the experimental features. She considered the coupling of an electronic triplet state to vibrational modes transforming as E_g and T_{2g}. For equally

Table 13. Stress and strain coupling coefficients for the CaO F^+ zero-phonon line (after Hughes and Runciman 1965).

Stress direction in measurement	Stress coupling coefficients, $cm^{-1}\ kg^{-1}\ mm^2$		
	A	B	C
[001]	0·40 ± 0·16	0·54 ± 0·09	—
[111]	0·42 ± 0·11	—	0·87 ± 0·19
[110]	0·54 ± 0·12	0·61 ± 0·14	1·00 ± 0·28
Average	0·45	0·58	0·94
	Strain coupling coefficients, $10^4\ cm^{-1}$		
	A'	B'	C'
	1·53	0·48	0·77

strong coupling to these modes the quenching factors are $K(E) = K(T_{2g}) = \frac{2}{5}$. Hughes (1970) showed that this model satisfactorily explains the absorption results when the cubic modes, A_{1g}, are less strongly coupled than the non-cubic modes. Thus there is no inconsistency in the strain coupling coefficients being reduced by only $\frac{2}{5}$ whilst the spin-orbit coupling reduction factor is of order 0·05. The complex Jahn–Teller model invalidates the equations developed in § 2.2.1: a comparison of the appropriate relationships at 0°K is shown in table 14. Obviously the values of S obtained using the relationships for P_{00} and Δ for both models are identical. However use of the single mode model with measurements of M_2, the second moment of the broad band, will overestimate S because of the factor 2·5, which expresses the strength of coupling to non-cubic modes, being absent. The two mode model involving equal coupling to E_g and T_{2g} modes also provides a basis for understanding the non-Gaussian band shape in fig. 25. This shape is to some extent sample dependent and great care must be exercised in evaluating the moments of the band. However there are evidently vestiges of structure in the broad band suggesting that

Table 14. Comparison of breathing mode and double Jahn–Teller coupling models

	P_{00}*	Δ*	M_2*
Breathing mode	e^{-S}	$S\hbar\omega$	$S\hbar^2\omega^2$
Double Jahn–Teller†	$e^{-(S_A+S_{NC})}$	$(S_A+S_{NC})\hbar\omega$	$(S_A+2·5S_{NC})\hbar^2\omega^2$

* P_{00} is the ratio of areas under zero-phonon line and broad band, Δ the separation of zero-phonon line and broad band peak, M_2 the second moment of broad band.
† Expressions derived assuming $S_E = S_{T_2} = S_{NC}$ and $\omega_A = \omega_E = \omega_{T_2} = \omega$.

the band is triple-peaked. Using a semi-classical model Cho (1968) has shown that a triple peaked band is expected when E_g and T_{2g} modes of vibration are coupled to a T_{1u} orbital state. Hughes (1970) shows that the band shape is reflected in the ratio of the fourth moment M_4 to the square of the second moment M_2. For equal coupling to E_g and T_{2g} modes M_4/M_2^2 is 2·1 compared with a value of 3·0 for a Gaussian band resulting from coupling to A_{1g} or E_g modes alone. The measured values for the F^+ band in CaO are ranged about 2·3–2·4 (Bessent and Hall 1971, Escribe and Hughes 1971, Henderson et al. 1972). Calculated quantum mechanically, the bandshape is conspicuously similar to that obtained experimentally (O'Brien 1971). Values of the Huang-Rhys factors vary according to the manner in which they are measured. Using a cluster model to represent the F^+ centre Hughes (1970) finds $S_A = 1·8$, $S_E = S_T \approx 4$ from the strain coupling coefficients. A later study of the temperature dependence of the bandshape by Escribe and Hughes (1971) yielded $S_E = S_T = 3·3$, $S_A = 1·6$, $\omega_E = \omega_T = 300 \text{ cm}^{-1}$, $\omega_A = 200 \text{ cm}^{-1}$: the measurement of Henderson et al. (1972) are in good agreement with these values.

Merle d'Aubigné and his colleagues have made conspicuous contributions to this problem. They measured the spin orbit coupling constant using magnetic circular dichroism (Merle d'Aubigné and Roussel 1971): The unquenched value of λ' pertinent to the broad band is $\lambda' = 31 \pm 6 \text{ cm}^{-1}$ whereas they found $\lambda' = -0·58 \pm 0·06 \text{ cm}^{-1}$ from studies of the zero-phonon line. Thus the Ham reduction factor has a value $K(T_1) = 0·02$. By careful measurements of the second and third moments of the dichroism they find $S_E = S_T = 3·15$ and $S_A = 2·05$ assuming equal coupling to E_g and T_{2g} modes and $\omega = 280 \text{ cm}^{-1}$ for all modes. Duran, Merle d'Aubigné and Romestain (1972) have measured the linear dichroism of the F^+ band and zero-phonon line under stress. They determine the following values for the reduction factors; $K(E) = 0.40$ and $K(T_2) = 0.39$, demonstrating clearly that this system is an excellent example of the special type of Jahn–Teller effect considered by O'Brien (1969). Romestain and Merle d'Aubigné (1971) have used group theoretical techniques to deduce a simple form of the general vibronic wave function assuming only that equal coupling to E_g and T_{2g} modes with equal energy introduces extra symmetry. From this they are able to calculate values of $K(T_1)$, $K(E)$ and $K(T_{2g})$, in excellent accord with the measured values.

No similarly detailed measurements of band shapes have been reported for the other oxides. However a start has been made in SrO. Hughes and Webb (1973) have measured the shape and moments between 4·2 and 400°K. They find an effective frequency of the vibrational modes interacting with the centre $\hbar\omega = 236 \pm 20 \, cm^{-1}$. In addition it is observed that as in CaO the coupling is stronger to non-cubic modes than to cubic modes. However since $S_{NC} = 4 \pm 1$ and $S_A = 3 \pm 1$ are larger than they are in CaO no zero-phonon line is observed. The coupling is less strong than in MgO where $S > 30$.

5.4. *Luminescence of F^+ centres*

Once the position of the F^+ absorption band was assured it was natural to look for the luminescence spectrum. So far F^+ centres have been observed in emission in MgO, CaO and SrO. In MgO the F^+ luminescence may be excited either by pumping in the 4·95 eV absorption band or by X-irradiation (Chen et al. (1969), Kappers et al. (1970)). There is a complication in that when both F and F^+ centres are present there may be overlapping emission due to both centres. However by varying the concentrations of F and F^+ centres (the latter being monitored by ESR), it may be shown that the F^+ luminescence band peaks at 3·13 eV. The band half-width is 0·6 eV at 77°K, which is very close to that of the absorption band. The shape of the luminescence when excited by pumping in the F^+ band is strongly sample and wave length dependent. For example in samples containing up to 3×10^{18} F^+ and F centres one may observe one, two or three overlapping bands depending upon the wave length of the exciting light (Cibert, Edel and Henderson, to be published).

The identification of the F^+ emission band in CaO was not complicated by overlap of the F and F^+ absorption bands, indeed they are well separated. Figure 25 shows that the emission band also has an attendant fine structure at low temperature. However the emission band is in no sense a mirror image of the absorption band; the flat-topped appearance so obvious in absorption is totally absent in emission. Indeed the emission band has a smaller half-width and is more nearly Gaussian, a reflection on the orbital singlet character of ground state. The vibronic sidebands are similar in position for both absorption and emission bands, the most prominent components being displaced from the zero-phonon line by $\sim 205 \, cm^{-1}$, $302 \, cm^{-1}$ and $344 \, cm^{-1}$ (Henderson et al. 1969, 1972, Evans and Kemp 1970).

Experiments on the stress-induced linear dichroism have been utilized by Hughes et al. (1972) and Duran et al. (1972) to determine the symmetry of the participating vibrational modes. The dichroism in both the zero-phonon line and the vibronic structure is typical of a $T_{1u} \rightarrow A_{1g}$ transition, when allowance is made for the sample depolarization due to internal strains. Although rather broad the trends in the $205 \, cm^{-1}$ sideband are the same as in the zero-phonon line and consequently this sideband is predominantly an A_{1g} mode. The sharp peak at $302 \, cm^{-1}$ has more or

Fig. 26

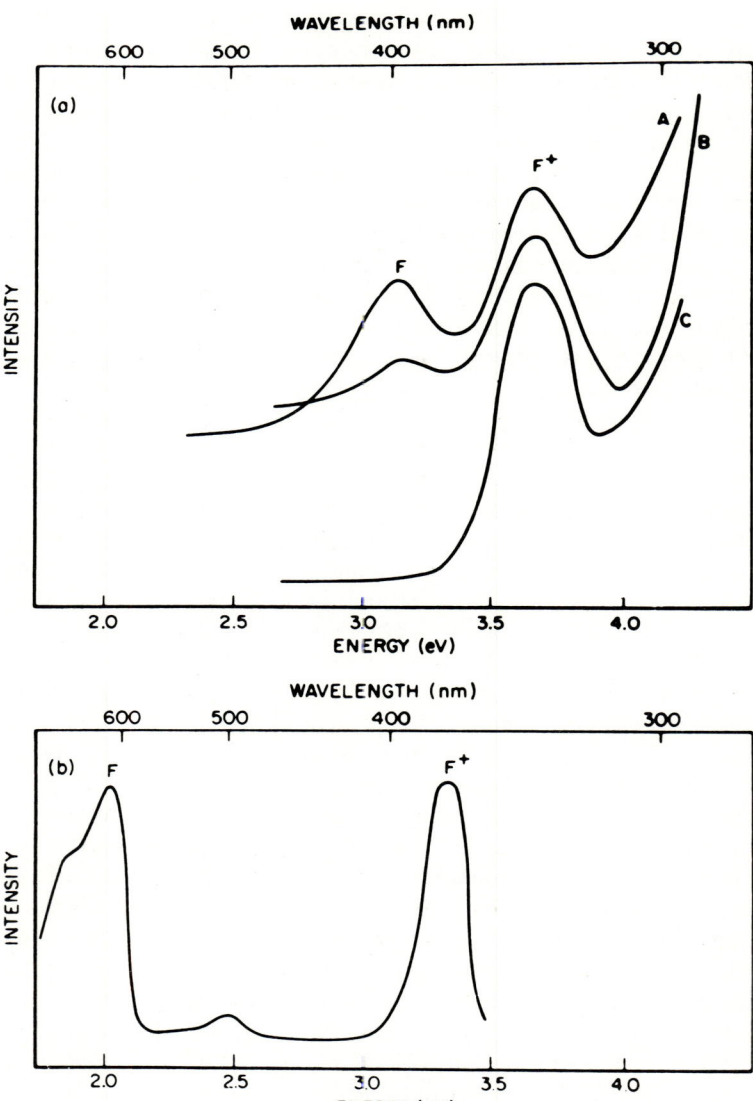

(a) F^+ and F absorption bands in CaO at 300°K. Curves A and B refer to two different samples taken from the same additively coloured crystal and curve C is for a neutron irradiated sample.
(b) Emission spectra from an additively coloured crystal at 300°K. Excitation in the F absorption band excites only the emission bands at 2·5 and 2·0 eV; the 2·5 eV band is due to unidentified impurities. The F^+ emission band is excited only by absorption in the F^+ absorption band. (After Henderson *et al.* 1969.)

less the same intensity for [100], [110] and [111] stresses, i.e. there is relatively little dichroism. Such an effect is expected when there are roughly equal contributions from E_g and T_{2g} modes since E_g and T_{2g} components polarize in opposite directions. The E_g and T_{2g} components of the σ-polarized spectrum are separated by 13–15 cm^{-1}. Note should be taken of the peak in the pure lattice density of states near 205 cm^{-1} (Saunderson and Peckham 1971). This peak corresponds to $TA(X)$ and $TA(L)$ critical points which give no A_{1g} modes, only modes of E_g and T_{2g} symmetry (Loudon 1964). Presumably this peak is due to a resonant mode. There is no similar peak in the density of states curve at 302 cm^{-1}, so that this sideband presumably a composite of T_{2g} and E_g modes. It is comforting that there is nothing in the emission results which conflicts with the two mode theory based initially upon the absorption experiments.

Less detailed studies have been made on F^+ centre emission in SrO. Hughes and Webb (1973) report bands at ~ 2.71 eV, ~ 2.48 eV and ~ 2.24 eV under excitation in F^+ absorption band at 77°K. Evans and Kemp (1970) have assigned to the F^+ centre the 2.48 eV peak.

It should be noted that the emission band positions may be predicted from knowledge of the electron-lattice coupling constants. For example in MgO and F^+ emission is predicted to occur at 2.4 eV (Henderson et al. 1968): the comparable value for CaO is 2.5 eV, neglecting the stronger coupling to non-cubic modes. When allowance is made for coupling to E_g and T_{2g} modes, in O'Brien's model (1971) the Stokes shift is given by $\delta = 2(S_A + S_{NC})\hbar\omega$. Hence we obtain emission energies of 3.9 eV, 3.3 eV and 2.73 eV for F^+ centres in MgO, CaO and SrO respectively.

5.5. F centres

We have already noted that anion vacancies may trap two electrons: in consequence diamagnetic F centres are produced. It is well established that both F^+ and F centres may be present together in MgO and CaO: indeed optically induced $F \leftrightarrow F^+$ conversion has been studied in both oxides as noted in Chapter 1. Insofar as the F centre is represented by two electrons trapped in a cage of positive charge it may be regarded as in 'inside out' helium atom. In cubic symmetry the ground state is $^1A_{1g}$; singlet excited states of both A and T symmetry may also be formed. The only allowed transition is $^1A_{1g} \rightarrow {}^1T_{1u}$, and this has been identified in MgO and CaO. At least in MgO the identification is based largely upon the photoconversion process: even here there are inherent difficulties because of the overlap between F and F^+ bands noted earlier. Chen et al. (1969) were able to decompose the composite F/F^+ band at room temperature into an F^+-band centred at 4.90 eV and an F-band at 5.01 eV with half-width 0.77 eV and oscillator strength $f = 1.25$. This f-value contrasts with that determined by Kappers et al. (1970) who find $f = 0.77$. Such a discrepancy is unacceptably large and requires further investigation as does the related discrepancy in the F^+ band half-width.

The situation is comparatively simple in CaO since the F^+ and F absorption bands are clearly resolved (fig. 26); initially the evidence that the band at 3.7 eV was the F^+ band with the F-band at 3.1 eV also relied on the photoconversion (Kemp et al. 1967 b). Later work on the luminescence excited by F-band absorption has amply confirmed the original assignment (Henderson et al. 1969, 1972). However, Modine (1972) provided the most crucial test in magnetic circular dichroism studies which clearly showed the states involved in the 3.1 eV band to have the appropriate diamagnetic character. The orbital g-value was measured as -0.34 ± 0.1 for the $^1T_{1u}$ state.

In SrO and BaO less work has been reported, although in the former F centres apparently absorb at 2·5 eV (Johnson and Hensley 1969). Chen *et al.* (1976) have investigated the 2·0 eV band which has previously been observed in both neutron and proton irradiated BaO as well as in additively coloured crystals (Bessent *et al.* 1968, Rose and Hensley 1972). This band has a strongly asymmetric shape which has been suggested to arise partially out of the overlap of F and F^+ absorption processes. Modine *et al.* (1976) conclude that this band is due entirely to F^+ centres.

A major interest in the F centres arises from the possibility of triplet states being excited optically. As noted above in the ground state the helium-like configuration ^1S transforms as the irreducible representation $^1A_{1g}$ in O_h symmetry. Excited configurations $(1s)(2s)$ and $(1s)(2p)$ will form singlet and triplet states ^1S, ^3S and ^1P, ^3P respectively. These atomic states in the crystal environment form the representations $^1A_{1g}, ^3A_{2u}, ^1T_{1u}$ and $^3T_{1u}$. Transitions to the $^1A_{1g}$ and $^3A_{2u}$ states are parity-forbidden and have not been observed. As discussed above the $^1A_{1g} \rightarrow ^1T_{1u}$ transition has been identified in MgO, CaO and SrO. The corresponding transition in emission has not been identified, although a fairly large Stokes shift is expected. Henderson *et al.* (1969) observed a weak band at 2·5 eV when exciting in the CaO 3·1 eV band, which they interpreted as the $^1T_{1u} \rightarrow ^1A_{1g}$ emission. Subsequent work shows that this emission band is not observed in all crystals so that it is apparently due to an unknown impurity. However a band at 2·0 eV, also shown in fig. 26, was found to have a long lifetime of 3 ms at 4°K, consistent with an oscillator strength of 5×10^{-7}. This f value is in line with that expected for a spin-forbidden transition. This led Henderson *et al.* (1969) to suggest that the 2·0 eV band is due to F band luminescence due to the $^3T_{1u} \rightarrow ^1A_{1g}$ transition. The band shows resolved structure at low temperatures, including a sharp phonon-less line at 5740 Å with phonon sidebands displaced to lower energies. Approximately the same phonon frequencies are involved in the vibronic structure as in the F^+ centre sidebands.

Using this fluorescence as the detecting light Edel *et al.* (1972) have observed EPR in the excited triplet state, confirming the earlier assignment. The spectrum portrayed in fig. 27 has been shown to be due to a triplet state in a tetragonally distorted symmetry site. Edel *et al.* (1972) attribute the distortion to a strong coupling of T_{1u} states to vibrational modes of E_g symmetry; this leaves the orbital component $|0\rangle$ lowest and it is from this level that no emission occurs. The Hamiltonian for this two electron centre should be written

$$\mathcal{H} = \lambda(l_1 \cdot s_1 + l_2 \cdot s_2) + g_L \mu_B \vec{B} \cdot (\vec{l}_1 + \vec{l}_2) + g\mu_B \vec{B} \cdot (\vec{s}_1 + \vec{s}_2) + \mathcal{H}_d \quad (46)$$

where $\mathcal{H}_d$ represents the dipolar coupling between the two spins, g_L is the orbital g-factor and $g = 2.0023$. Second order effects in $\vec{l} \cdot \vec{s}$ and the orbital Zeeman interaction give the following expressions

$$g_\parallel = 2.00, \qquad g_\perp = g_0\left[1 - \left(\frac{g_L}{g_0}\right)\frac{\lambda}{3E_{JT}}\right] \quad (47)$$

and

$$D = D_d + \frac{\lambda^2}{12 E_{JT}} - \frac{\lambda'^2}{4\Delta}. \quad (48)$$

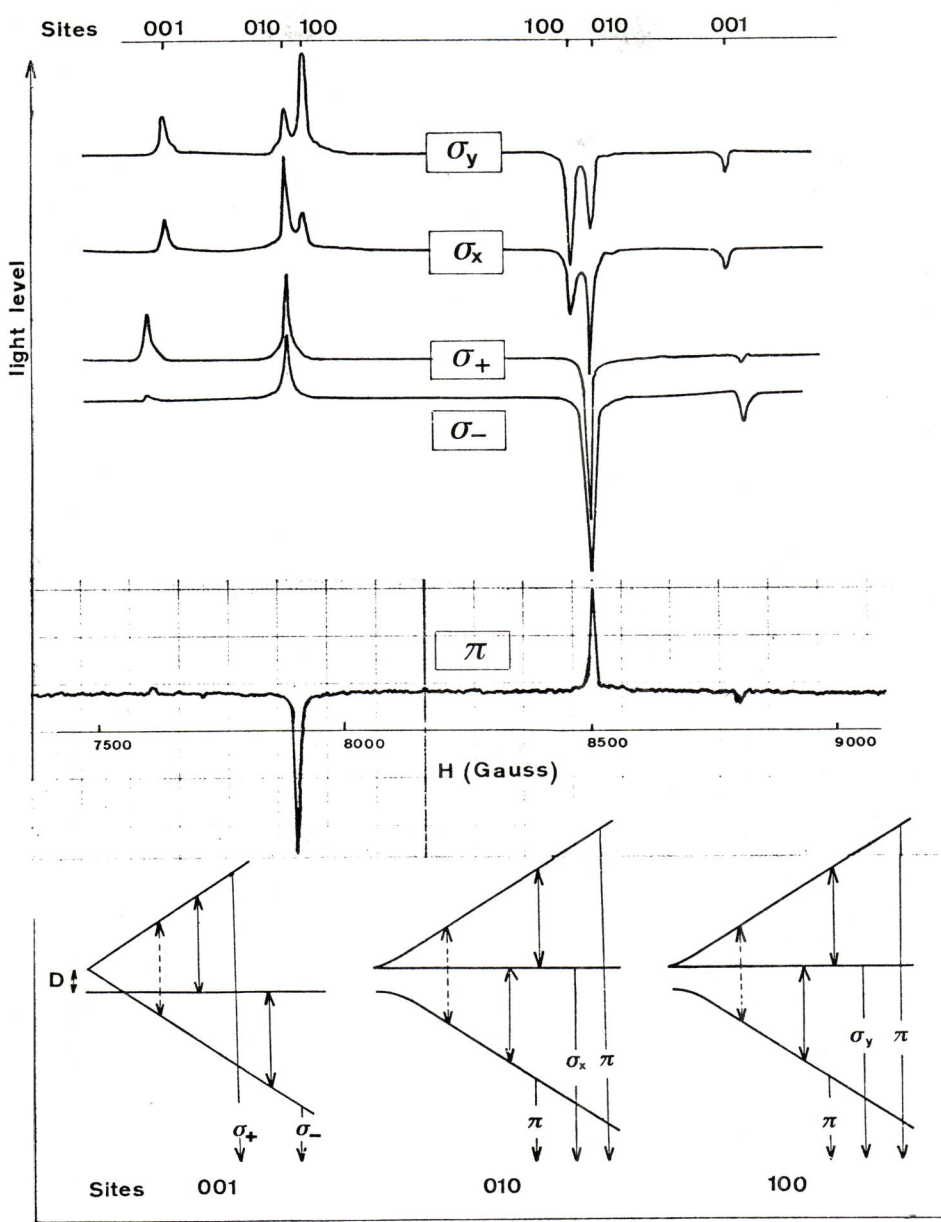

ESR in the optically excited ^3P state of the F centre in CaO using the intensity variation of emitted light of various polarizations. The magnetic field is parallel to a $\langle 100 \rangle$ axis, except for σ_x and σ_y where a small misalignment was used to show the two-fold coincidence of lines. In (c) the predicted selection rules for the emission of polarized light are shown (after Edel *et al.* 1972).

In these equations E_{JT} is the orbital splitting of the $^3T_{1u}$ level, Δ is the energy difference between $^3T_{1u}(|z\rangle)$ and $^1T_{1u}(|x\rangle, |y\rangle)$ and the λ's represent spin–orbit coupling in the $^3T_{1u}$ and $^1T_{1u}$ levels. From the orientation dependence of the spectrum one deduces $g_\parallel = 1.9991$, $g_\perp = 1.998$ and $D = 563 \times 10^{-4}$ cm^{-1}. From $g_\perp$ one deduces $g_\perp \lambda / 3E_{JT} \approx 2 \times 10^{-3}$. Furthermore since the triplet → singlet transition is made partially allowed by spin–orbit mixing the oscillator strength $f \approx (\lambda'/2\Delta)^2$: using the f value measured by Henderson et al. (1969) $\lambda'/\Delta = 1.4 \times 10^{-3}$. Since $3E_{JT} \approx 2500$ cm^{-1} (Edel et al. 1972) we may deduce that the product $g_L \lambda \approx 5$ cm^{-1}. Unfortunately λ', λ and Δ have not been determined independently. The maximum value of Δ is given by the energy difference between the peak positions of absorption and emission bands, i.e. ~ 8900 cm^{-1}. Thus we estimate that $(\lambda')^2/4\Delta \approx 0.004$ cm^{-1} or less than 10% of the measured value of D. Assuming $\lambda' \approx \lambda$ then there is a further contribution $\lambda^2/12E_{JT}$ of order 0.014 cm^{-1}. In consequence it is safe to conclude that second order spin–orbit admixtures contribute not more than $\sim 20\%$ of D and one reasonably attributes D mainly to spin–spin dipolar interaction.

Since the centre is distorted tetragonally there must be a removal of orbital degeneracy either by Jahn–Teller coupling to vibrational modes of E_g symmetry or by static crystal field effects. The latter possibility is ruled out by the spectra in fig. 27, which indicates the nature of the selection rules for triplet to singlet transitions. Consider defects distorted along [001], this crystal direction being parallel to the magnetic field. The only non-zero matrix elements of the angular momentum are of the form $\langle p_x | l_z | p_y \rangle$ which couples $^3T_{1u}(z)$ to $^1T_{1u}(x, y)$. Consequently the emitted light is polarized along the x or y directions. Thus for [001] defects with $\vec{B}$ parallel to [001] the $|+1\rangle$ and $|-1\rangle$ states emit σ_+ and σ_- light respectively and the $|0\rangle$ state does not radiate. However [100] and [010] sites which are both perpendicular to $\vec{B}$ emit light which is σ_x, σ_y or π-polarized. It is in keeping with these selection rules that the extreme line at high field and at low field is not observed when detecting π-polarized light, since they arise at the [001] site. Furthermore D is positive since there is an increase in σ_+ intensity when the low-field line is partially saturated (i.e. the $|0\rangle \rightarrow |+1\rangle$ transition, whereas when the high field line is partially saturated there is an increase in σ-intensity due to the decreased population of the $|-1\rangle$ level. In addition the spectra in fig. 27 show that the selection rules are not perfectly obeyed. This is interpreted as due to tunnelling between three different vibronic states of the same centre in a time short relative to the radiative lifetime (Edel et al. 1972). Observation of dynamical effects has been discussed in rather more detail by Merle d'Aubigné (1974).

Edel et al. (1974) have also studied the zero-phonon line of the F centre luminescence band under uniaxial stress. That no splitting is observed when the stress is applied parallel to the [111] axis is consistent with the excited level being coupled mainly to E_g vibrational modes. These experiments also show that the ratio of tunnelling time to radiation lifetime, τ_N/τ is less than 2.5×10^{-3}.

No analogous results have as yet been reported for F centres in MgO or SrO.

5.6. Theoretical studies of F and F^+ centre properties

The simplest model of the F^+ centre in an oxide MO assumes that the interactions of the trapped electron are confined to the six nearest-neighbour cations. For MgO the assumption that the wave function is a linear combination of $3s$ and $3p$ orbitals of adjacent Mg^{1+} ions leads to half the observed anisotropic splitting (Wertz et al. 1957). The coefficients of the $3s$ and $3p$ orbitals are chosen so as to give the observed isotropic splitting. The latter corresponds to a magnitude of 0.276×10^{24} cm^{-3} = $|\psi(Mg)|^2$, as contrasted with the value $|\psi(Mg^{1+})|^2 = 17.1 \times 10^{24}$ cm^{-3} for the free ion (Crawford et al. 1949).

A more careful approach to the understanding of the F^+ centre in MgO takes into account the possibility of using the 3s and 3p functions of the next-nearest cations (Kemp and Neeley 1963 a). There are then a number of independent LCAO functions for each representation of the full octahedral group O_h. The possible 3s, 3p and also 3d LCAO functions for nearest and next-nearest-neighbours cations have been tabulated. These are then, in principle, to be used with the point-ion-lattice Hamiltonian (Gourary and Adrian 1957):

$$\mathcal{H} = \frac{-1}{2r^2}\frac{d}{dr}\left(r^2\frac{d}{dr}\right) + \frac{l(l+1)}{2r^2} + V_0(r). \tag{49}$$

In the actual calculation the wave functions employed were approximated by a single spherical harmonic term corresponding to the lowest value of l appropriate to the function expressed in cubic harmonics (Bethe and von der Lage 1947). Linear combinations of these truncated functions are then used in the point-ion lattice Hamiltonian H. The secular equation is obtained by minimizing the expectation value of H with respect to the coefficients in the linear combination wave function.

The ground state is presumed to be represented by the s-like function Γ_1, while the first excited state is represented by Γ'_4, using only the nearest-cation 3s and 3p functions. Slater functions were used for the Γ_1 and Γ'_4. The energies obtained for Γ_1 and Γ'_4 are:

$$\Gamma_1 : E_0 = -19.1 \text{ eV}, \qquad \Gamma'_4 : E_1 = -14.4 \text{ eV}.$$

A large polarization correction for the 12 adjacent O^{2-} ions and a smaller distortion due to outward displacement of the six magnesium ions are in order. The positions of all ions are held at their normal lattice spacings as an electron is brought from infinity. Subsequently, the nearest-cation neighbours are allowed to relax outwards. With zero distortion but with polarization, the corrected energies are given by:

$$E_0 = -15.2 \text{ eV}, \qquad E_1 = -10.5 \text{ eV}.$$

Allowing a 5% distortion outwards for the six nearest-cation neighbours, the energies become:

$$E_0 = -12.9 \text{ eV}, \qquad E_1 = -8.2 \text{ eV}.$$

Remarkably, despite the large corrections to E_0 and to E_1, the transition energy has remained unchanged. This magnitude for the fractional distortion is in agreement with data for MgO and SrO, which indicates a distortion between 4 and 8% (Unruh and Culvahouse 1967, Culvahouse et al. 1965).

It is noteworthy that even with the highest value of E_0 the calculations suggest that the ground state lies *below* the maximum of the oxygen 2p valence band. This assumes a value of 8.7 eV for the band gap in MgO (Reiling and Hensley 1958). To avoid the difficulty of strong coupling of the F^+ centre to band functions which would be inconsistent with the observed localization, it is presumed that the valence band is depressed over several lattice constants. The remarkable corollary is that the F^+ band in MgO cannot be bleached by ionizing radiation; i.e. such radiation will not remove the F^+-centre electron from the vacancy. However, the $F^+ \leftrightarrow F$ conversions are not precluded.

Agreement of the 4·95 eV band position established for the F^+ centre in MgO by correlated optical and ESR experiments with the predicted value of 4·7 eV must be regarded as remarkably good. The estimated values of F^+ bands in CaO, SrO and BaO are respectively 3·8, 3·4 and 3·0 eV (Kemp 1963). The experimental values have been given earlier as 3·7, 3·0 and 2·0 eV for the same oxides. The excellent accord for CaO and SrO, combined with the very good prediction for MgO, must be regarded as a triumph, whatever the approximations involved. The prediction for BaO is still good enough to be meaningful in implicating the 2·0 eV band. For BaO the polarization or the distortion corrections may be in error, as may the neglect of the finite ion sizes in the case of BaO.

The deviations of the F^+-centre g factors (table 5) from the free-spin value are negative for all of the oxides (except MgO, for which $\Delta g = 0$), just as for the alkali halides. Accepting the deBoer model and making use of the 3s and 3p orbitals of Mg^{1+} to calculate Δg by a modification of an approach due to Adrian (1957) leads necessarily to a negative deviation (Kemp and Neeley 1963 a). An attempt has been made to devise a theory of g factors which would apply to F^+ centres in alkaline earth sulphides, selenides and tellurides in addition to oxides. For many of these compounds, Δg is observed to be positive. A positive contribution to the g factor of all F^+ centres (including oxides) is obtained by admixture of a configuration derived by transfer to the anion vacancy of a second electron; this comes from a nearest-neighbour anion. However, for F^+ centres in CaO, SrO and BaO, the magnitude of the negative contribution must be considerably larger than the positive one (Bartram et al. 1967, Kemp and Neeley 1963 a).

More recent theoretical studies have aimed at treating the electron–phonon coupling more satisfactorily. Wood and Öpik (1968, unpublished) obtain qualitative agreement with the F^+ band position. More significantly, in view of the apparent controversy over the F/F^+ absorption band was their conclusion that the F and F^+ absorption bands should be almost coincident. Experimentally this is now known to be the case. Bennett (1968, 1969 and 1970) has used several models to calculate absorption and emission energies, excited state lifetimes and Huang-Rhys factors for both F and F^+ centres in MgO, CaO and SrO. Unfortunately he included coupling only to the breathing modes of A_{1g} symmetry: experimentally the coupling to T_{2g} and E_g modes has been shown to be much the more important. O'Brien's (1971, 1976) quantum mechanical calculation of the band shape for F^+ absorption in CaO is in remarkable agreement with experiment, although it is less satisfactory in the prediction of the emission band shape (O'Brien 1976).

The theoretical study of F^+ centres has now been extended to the absorption and emission bands of F centres. Various calculations on F centres in MgO concur that the F and F^+ bands are in close proximity (Neeley 1964, Neeley and Kemp 1964, Bennett 1969, 1970, Wood and Öpik 1968, unpublished). There is also good agreement between the predicted and measured position of $^1A_{1g} \to {}^1T_{1u}$ absorption band and $^3T_{1u} \to {}^1T_{1u}$ emission bands. The most interesting work is that of Wood and Wilson (1975): these calculations, which apply to the F centre in CaO, used the same techniques as had been developed for F and U-centres in the alkali halides. They adopt a smoothly varying Slater-type orbital $\psi_F(r)$ normalized to the atomic orbitals ϕ on neighbouring ions:

$$\psi(r) = \psi_F(r) - \sum_{kl} \langle \psi_F | \phi_{k1} \rangle \phi_{k1}(r - R_k).$$

The summation is taken over the l atomic orbitals on the k ions of the inner

Fig. 28

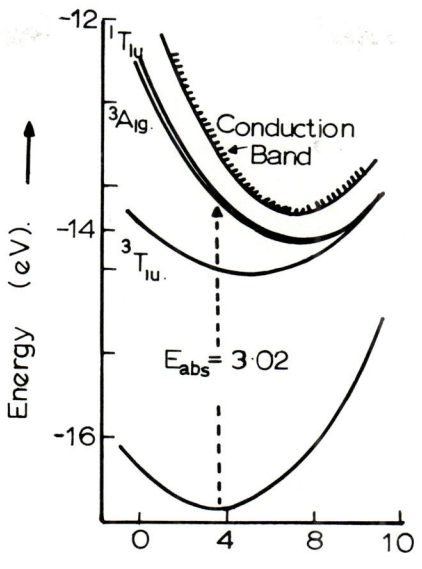

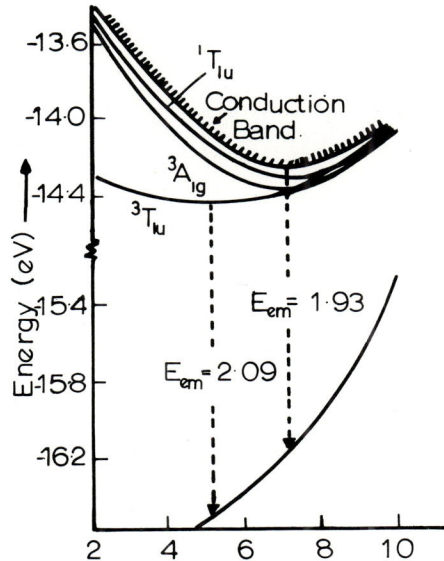

% Relaxation of nearest neighbour distance

Showing theoretical configurational coordinate curves for ground and excited states of F-centres in CaO. The energies and $^1A_{1g}$, $^3T_{1u}$, $^3A_{1g}$ and $^1T_{1u}$ have been calculated for A_{1g} and E_g displacements of first shell ions around the anion vacancy (after Wood and Wilson 1975).

region of the crystal around the defect. In the outer region a simple effective mass approximation is used. The Hamiltonian is a sum of two one-electron equations plus the electron–electron interaction. Configuration co-ordinate diagrams for $^1A_{1g}$, $^3T_{1u}$, $^3A_{1g}$ and $^1T_{1u}$ have been calculated for A_{1g} and E_g displacements of the nearest neighbour ions around the anion vacancy. The resulting energy level schemes appropriate to both the absorption and emission transitions are shown in fig. 28. The $^1A_{1g}$ and $^3T_{1u}$ states are strongly bound, the charge density of the two electrons being almost entirely restricted to the anion vacancy. In the $^3A_{1g}$ and $^1T_{1u}$ states the charge density is much more diffuse. F centre emission, according to these models should then occur at 2·09 eV [$^3T_{1u} \to {}^1A_{1g}$] and 1·93 eV [$^1T_{1u} \to {}^1A_{1g}$]. The former prediction is much to be expected from measurements at low temperatures (Henderson et al. 1969, 1972). At high temperature (320–370°K) it is observed that the $^3T_{1u} \to {}^1A_{1g}$ emission at $\sim 2\cdot 0$ eV is gradually replaced by a band at slightly higher energies. This is accompanied by a rapid decrease in the lifetime of the luminescence. Bates and Wood (1975) suggest that this behaviour is consistent with $^1T_{1u}$ being increasingly populated with increasing temperature such that the intensity is rapidly shifted to the $^1T_{1u} \to {}^1A_{1g}$ transition since this is allowed.

Wood and Wilson (1975) also find good agreement with the vibronic properties of the F centre. First they estimate values of $S = 12$ and $W(0) = 0\cdot 35$ eV for the absorption band. For the $^3T_{1u} \to {}^1A_{1g}$ emission they calculate the effective frequency of E_g modes to be 265 cm^{-1}, and S values of 1·4 for E_g modes and 1·2 for A_{1g} modes. The static Jahn–Teller effect is found to lower the $^3T_{1u}$ level by 0·045 eV and the splitting between $^3T_{1u}(z)$ and $^3T_{1u}(x,y)$ is 0·18 eV. However, the Jahn–Teller effect in the excited $^1T_{1u}$ level is negligible due to its very diffuse nature. This latter point was ignored by Edel et al. (1972), who assumed the Jahn–Teller effect in the $^1T_{1u}$ state to be more like that in the $^3T_{1u}$ state.

§ 6. The structure and properties of defect aggregates

There are many examples in the alkali halides and alkaline earth halides of centres which consist of groups of vacancies (Klick 1972, Stoneham and Hayes 1975). Such groups are produced by thermal annealing of radiation damage, optical bleaching or simply very extensive irradiation. A neutral aggregate consisting of nearest neighbour anion and cation vacancies is referred to as a vacancy pair—in the Sonder–Sibley (1972) notation this is a P-centre. Alternatively the term divacancy is retained to describe the situation in which two anion vacancies or two cation vacancies are located in neighbouring sites. Obviously this situation is easily extended to more complex defect centres. An important class of defect centres is that in which extrinsic defects and intrinsic defects are clustered. We have already discussed this situation insofar as Cr^{3+}, Fe^{3+}, Mn^{4+} ions are charge compensated by nearby cation vacancies (Chapter 3) as well as the case of positive holes localized on O^{2-} ions near to monovalent ions or cation vacancies with adjoining impurities (Chapter 4). In much the same way F and F^+ centres may be located next to impurity ions. This is especially well documented in the case of the F_A centres in the alkali halides (Lüty 1968). Similar centres have been observed in calcium oxide. Such defects will be reviewed first in this chapter: however the bulk of the discussion concerns the aggregation of defects consequent upon annealing specimens which have previously been irradiated with energetic particles. The most detailed results have been obtained for neutron-irradiated magnesium oxide and our discussion faithfully reflects this emphasis. Generally the effect of annealing is to reduce the total defect concentration via interstitial/vacancy recombination. However, near-neighbour vacancies may cluster together to form pairs as well as more complex defect aggregates. Thus in annealed crystals evidence has been adduced which indicates the presence of the oxide analogues of the F_2, F_3 and F_4 centres observed in the alkali halides. It is emphasized that the annealing temperatures required to produce aggregation depends very much on the oxide used. It has been reported that defect pairs produced by annealing magnesium oxide at 300°C are destroyed at this temperature in calcium oxide (Wertz et al. 1961).

There are several EPR spectra after neutron doses in excess of $\sim 5 \times 10^{18}$ nvt which have been interpreted as vacancy aggregate centres (Wertz et al. 1961, Tanimoto et al. 1965, Henderson 1966, 1976 and Hall 1975). Only those defects for which the model is supported by compelling EPR evidence will be discussed further.

6.1. F_A^+ centres in calcium oxide

Hughes and Pells (1972) first adduced evidence for the existence of F_A^+ centres in CaO: they reported optical studies of neutron-irradiated crystals of CaO doped with nominally 1% of MgO. The F_A^+ centre, an F^+ with a nearest neighbour Mg^{2+} ion replacing the Ca^{2+} ion, was reported to have an absorption band at 374 nm, a complementary emission band at 398 nm, both bands being accompanied by a zero-phonon line at 384·5 nm. The zero-phonon line is evidence of a weak electron-phonon coupling. Measurements of the spectra shown in fig. 29 are consistent with a Huang–Rhys factor of $S \approx 4$ and effective frequencies of $hw \approx 220 \text{ cm}^{-1}$. The band shapes are explained without recourse to dynamical distortion effects necessary

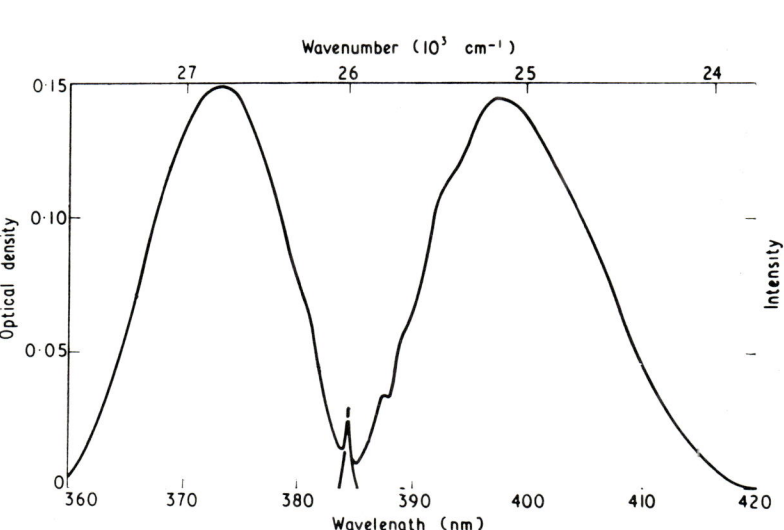

Fig. 29

The 374 nm and 398 nm emission bands for the F_A centre in CaO. No attempt was made to resolve fully all the sideband structure (after Hughes and Pells 1972). The absorption bandshape was better decomposed from overlapping bands (not shown) in subsequent work (see Boas et al. 1973).

in understanding the F^+ bandshapes in CaO (see Chapter 5). Application of uniaxial stress to the crystals results in splittings of the zero-phonon line which are linear in stress; they are consistent with the F_A^+ centre having tetragonal symmetry the transition moments being electric dipole in character parallel to the fourfold axis.

Of course F_A^+ centres are paramagnetic having an electronic spin $S = \frac{1}{2}$, consequently the crucial evidence to be obtained is the ESR spectrum. Weightman and Hall (1973) have observed three EPR spectra corresponding to defects with $S = \frac{1}{2}$ in sites of tetragonal symmetry. The B spectrum was characterized by $g_\parallel = 1.9996 \pm 0.0001$ and $g_\perp = 2.0006 \pm 0.0001$: subsequent observation of the hyperfine structure due to the 10% abundant ^{25}Mg nucleus confirmed these lines as due to the F_A^+ centre (Boas et al. 1973). These hyperfine lines, which were approximately one-fiftieth the intensity of the principal B lines, also show tetragonal symmetry with axes along the crystal $\langle 100 \rangle$ directions. From detailed studies of the orientation dependence of the hyperfine structure the following spin Hamiltonian parameters are deduced: $A_\parallel = -9.72 \pm 0.09$ MHz, $A_\perp = -7.50 \pm 0.06$ MHz and $|P| = 0.056 \pm 0.020$ MHz. The signs of $A_\parallel$ and $A_\perp$ where chosen in accordance with the known g_N value of ^{25}Mg and the measured signs for F^+ centres in MgO.

It is of interest to compare the hyperfine parameters of the F_A^+ centre in CaO with those of F^+ centres in both MgO and CaO (table 15). The Fermi contact inter-

Table 15. Hyperfine structure parameters of F^+ centres in MgO and CaO compared with those of F_A^+ centres in CaO

	a (MHz)	b (MHz)	P (MHz)	$|\psi_F(r)|^2$ 10^{-21} cm^3
MgO	−11.03	−1.33	+0.141	0.84
CaO	−25.66	−2.71	—	0.83
CaO:Mg	−8.24	−0.74	±0.056	0.88

action, a, is smaller for the F_A^+ centre than for F^+ centres, in keeping with the Mg^{2+} ion being rather further from the vacancy centre in CaO than in MgO. However, assuming $A_R = 325$ for Mg and $A_R = 700$ for Ca (Hughes and Henderson 1972) we find that $|\psi_F(r)|^2$ is roughly constant in the three cases. If the quadrupole interaction is interpreted in the same way as for F^+ centres (Chapter 5) then one obtains for the polarization contribution to the field gradient $q_P = 1.054/R^3$ with positive P and $q_P = -2.486/R^3$ with P negative. That these values are large relative to that obtained for F^+ centres in MgO is consistent with the Mg ion being displaced outwards by an amount which is greater than is the Mg ion displacement in F^+ centres in MgO.

Boas et al. (1973) also investigated the optical properties of the F_A^+ centre in greater detail. In addition to the band at 374 nm, which Hughes and Pells (1972) assigned to the $A \to A$ transition (F_{A1}^+ band) of the F_A^+ centre, Boas et al. (1973) observed a band at approximately the same wave length as that for the F^+ band. However two zero-phonon lines were attendant upon this band confirming it to be a composite band, as does the observation that the F^+ zero-phonon line is only one-third the intensity expected for such an intense band. Using this as a guide Boas et al. (1973) decomposed the band into two components, one due to the F^+ centre with the F_{A2}^+ band ($A \to E$ transition) at 337 nm. These conclusions are confirmed by subsequent polarized luminescence studies (Hughes and Pells 1975).

Other perturbed defects are detected by luminescence studies of the CaO:Mg crystals. Hughes and Pells (1975) have observed spectra due to an F_{4A}^+ centre comprising the linear array Mg–F^+–Mg aligned along the crystal $\langle 100 \rangle$ axes. Welch et al. (1976) have also detected luminescence due to F_A centres.

6.2. Exchange-coupled defect clusters

We have already discussed defects in which two holes are trapped on O^{2-} ions diametrically disposed across a cation vacancy. In such systems the electronic spins $s_1 = s_2 = \frac{1}{2}$ are coupled by the exchange interaction to give a new total spin angular momentum vector

$$\vec{S} = \vec{s}_1 + \vec{s}_2$$

where $S = 1, 0$. The analogous electron excess centres are exchange coupled F^+ centre pairs: the EPR spectra of such pairs have been reported in CaO (Tanimoto et al. 1965, Hall 1975) and MgO (Henderson 1966, 1976). The isotropic exchange interaction is

$$J\vec{s}_1 \cdot \vec{s}_2 = \tfrac{1}{2}J[S(S+1) - \tfrac{3}{2}]$$

which gives singlet and triplet states at $-3J/4$ and $J/4$ respectively. With $J > 0$ the singlet state lies lowest separated from the triplet state by J. With $J \gg$ crystal field effects EPR transitions within the triplet state manifold are described by a spin Hamiltonian

$$\mathcal{H} = g\mu_B \vec{B} \cdot \vec{S} + D_S[S_z^2 - \tfrac{1}{3}S(S+1)] + J\vec{s}_1 \cdot \vec{s}_2 \tag{50}$$

where $D_S = 3\alpha D_e + \beta D_c$ is the zero-field splitting of the spin state S written in terms of
  (i) D_c the zero field splitting of the uncoupled defects (in this case $D_c = 0$) and
  (ii) D_e a sum of dipolar interaction (D_d) and anisotropic exchange interaction D_E (negligible here).
Hence $D_S = \tfrac{3}{2}D_d$ since $\alpha = \tfrac{1}{2}$ for a spin system with $S = 1$, $s_1 = s_2 = \tfrac{1}{2}$.

In both MgO and CaO triplet state defects have been reported in room temperature measurements on neutron irradiated crystals. The spectra in arbitrary orientation consist of six equally intense lines, the orientation dependence of which (fig. 30) shows clearly the defect symmetry to be tetragonal with a fourfold rotation axis parallel to the crystal $\langle 100 \rangle$ direction. The values of D_d deduced from the orientation dependence studies show that the F^+ separation is $1 \cdot 07 \, a_0$ in MgO and $1 \cdot 14 \, a_0$ in CaO, where a_0 is the undistorted unit cell length. In consequence Tanimoto et al. (1965) proposed that the spectrum in CaO was due to a linear trivacancy in which a cation vacancy is interposed between two F^+ centres: the trivacancy axis coincides with the crystal $\langle 100 \rangle$ direction. The $^{25}Mg^{2-}$ hyperfine has been observed in association with the triplet state spectrum in MgO (Henderson 1966).

In both oxides measurements at low temperatures reveal the spin singlet to lie lowest. Thus the coupling is antiferromagnetic. Detailed studies of the temperature dependence of these $S = 1$ spectra (fig. 31) give the values of J shown in table 16. Theoretical estimates of J convincingly demonstrate that the cation site between the two F^+ centres must be vacant (Norgett 1971, Berezin 1969, 1972).

Below $10\,°$K these spectra are not observed since all defects are in the singlet state. The triplet state may be repopulated by optical pumping in the neighbourhood of the F^+ band. Tanimoto et al. (1965) first showed that even using broad band unpolarized light the optically excited spectrum showed unmistakeable evidence of spin-polarization due to level inversion. Extension of these studies using narrow band, polarized light revealed the details of the spin-selection rules during the optical

Fig. 30

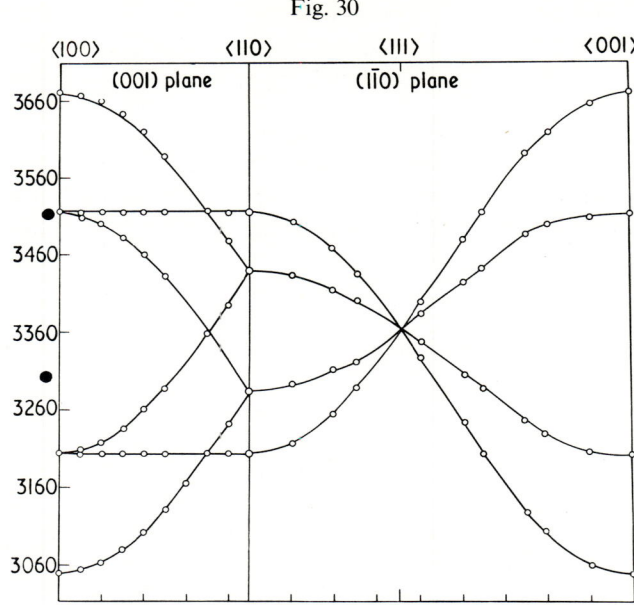

Portraying the orientation dependence of the ESR spectrum due to the $S = 1$ state of F^+ centre pairs in MgO. The spectrum was recorded at 300°K and X-band frequencies. (After Henderson 1966.)

pumping cycle (Henderson 1976). Of more importance however, the use of polarized light allows one to plot out the absorption spectrum of these aggregate centres, even though the absorption due to isolated F^+ centres is many times stronger than that due to the F^+ centre pairs. As can be seen in fig. 32 the absorption spectrum, plotted

Table 16. Spin Hamiltonian Parameters for F^+ centre pairs in MgO and CaO

Orientation	Crystal	g	$D_d(\text{cm}^{-1})$ Expt.	J	Reference
[100]	MgO	2·003	0·0191	78	Henderson 1966, 1976
[100]	CaO	2·001	0·0105	55	Tanimoto *et al.* 1965
[100]	CaO	2·002 (∥) 1·9985 (⊥)	·0248	960	Hall* 1975
[110]	CaO	2·001 (∥) 1·9999 (⊥)	80	150	
[121]	CaO	1·9990 (∥) 2·006 (⊥)	·0114	130	

* The spectra do not have simple ⟨100⟩ symmetry axes. The lower symmetry revealed by EPR studies makes assignment of models rather more speculative than the tetragonally symmetric defects. In consequence they are not further discussed in the text.

Fig. 31

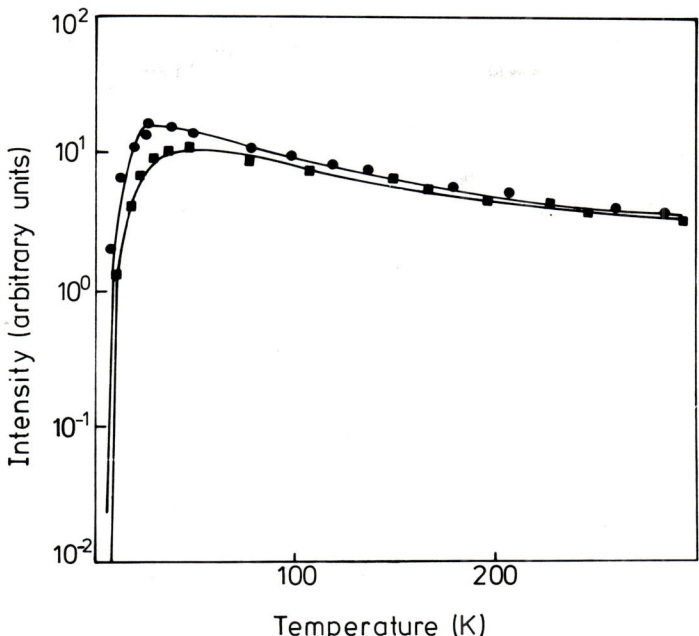

Temperature dependence of $S = 1$ state ESR spectrum due to F^+ centre pairs in MgO and CaO. (After Henderson 1976.)

Fig. 32

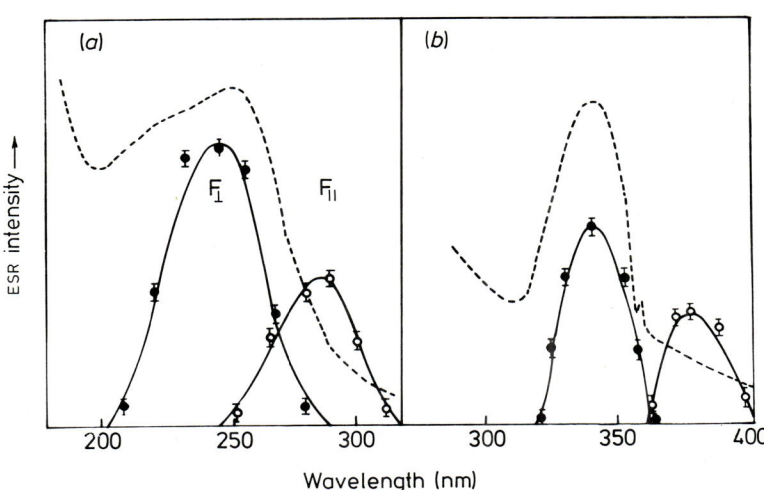

Absorption properties due to F^+ centre pairs in MgO obtained by recording the optically pumped ESR spectrum as a function of exciting wavelength at low temperature. (After Henderson 1976.)

by monitoring the pair ESR spectrum as a function of exciting wave length, shows two bands in obvious similarity with the F_A^+ centres discussed above.

6.3. *Magnetic resonance studies of annealed crystals*

Heating crystals of the oxides after neutron irradiation reduces the amplitude of the F^+ centre lines. This reduction is partially attributable to the formation of P^- centres by aggregation between the F^+ centres and cation vacancies. Recombination processes as well as the formation of more complex aggregate centres are also involved. As a tentative model for the P^- centre, Wertz et al. (1961) postulated that the experimental evidence was consistent with an electron trapped at an anion-cation vacancy pair. This defect was first detected using paramagnetic resonance: a single line spectrum shifted from the F^+ centre line in magnesium oxide by $\Delta g = -0.0015$ was observed. Larger g shifts were detected for the heavier oxides. The line was asymmetric in shape, the asymmetry being dependent upon the orientation of the crystal relative to the Zeeman field. In magnesium oxide the P^- centres are extremely stable since they are present even after heating to temperatures of 900°C. After this treatment the intensity of the P^- line is considerably enhanced by X-irradiation: No F^+ centres survived this treatment. P^- centres have also been detected in MgS, SrS, SrSe, and BaSe (Wertz et al. 1961).

As in the F^+ centre, the electron trapped in the P^- centre is largely in an s-like ground state. Thus the non-cubic component of the crystalline field introduced by the neighbouring cation vacancy will have only a small effect and only a small essentially isotropic shift in g value is expected. The appearance of the lines at X band (9.5 GHz) essentially conform with this view. However, the electron will spend some of its time in the orbitals of the neighbouring cations so that it is never in a pure s state. The contributions from non-s orbitals will shift the g value and this shift will be accentuated by two other contributions. Because of the missing cation, the electron is attracted away from the centre of the negative ion vacancy and towards that cation opposite the cation vacancy. The electric field resulting from the vacancy will also polarize the cation more than in the F^+ centre. Both these effects increase the amount of p and d character admixed in the P^- centre wave function and increase the departure of the g value relative to the F^+ centre. The small contributions from p and d states will be affected by the axial component of the crystal field and consequently $g_\perp \neq g_\parallel$. In general the differences $g_\perp - g_\parallel$ are not easily resolved at X band but may be resolved at Q band (To et al. 1969). The values of $g_\parallel$ and $g_\perp$ are given in table 7 for comparison with the average g values determined at X band and the g values of surface F^+ centres in powders (S centres) (Nelson et al. 1967).

No detailed theory is yet available for the g values of the P^- centre although one might expect some qualitative similarities with the simple F^+ centre. However, the positive g shift from configuration mixing with O^- charge transference states (Bartram et al. 1967) should be negligible since the P^- centre is negatively charged and will tend to repel electrons from the near-neighbour anions. In addition the electron is likely to be less strongly bound to the dipole than is the electron in F^+ centres and consequently contributions from admixtures with O^{3-} charge transference configurations should be included. However, although such contributions give a positive g shift, it is expected to be negligibly small for the reasons given by Wertz et al. (1961) in the case of the F^+ centre. The effective spin–orbit coupling in the F^+ centre increases monotonically along the series MgO to BaO. Making the

Table 17. Some parameters for the P^- and S (surface) centres in the alkaline earth oxides

Host	E_B (calc)† eV	E_B (obs.) eV	P^- centres‡			S centres§		
			g_{av}	$g_\parallel$	$g_\perp$	g_{av}	$g_\parallel$	$g_\perp$
MgO	2·42	3·6	2·0007	2·0004	2·0012	2·0007	2·0016	2·0003
CaO	1·72		1·9995	1·9995	1·9980	1·9977	1·9992	1·9969
SrO	1·70		1·9816	1·9839	1·9804	1·9798	1·9846	1·9792
BaO	1·24							

† To, Stoneham and Henderson 1969.
‡ g_{av} values measured at X-band (Wertz et al. 1961), $g_\perp$ and $g_\parallel$ values measured at Q band (Henderson, unpublished).
§ X-band measurements (Nelson et al. 1967).

reasonable assumption that such a situation obtains also in the P^- centre the increasing shift to smaller g values along the series is to be expected. It is not obvious why $g_\perp > g_\parallel$ in the case of magnesium oxide.

The hyperfine structure of the P^- centre is expected to also show interesting differences from the simple F^+ centre. As an example we consider only the P^- centre in magnesium oxide in detail. For the P^- centre there are only five nearest-neighbour cations compared with six cations in the F^+ centre. The absence of one positively charged ion will cause the electron to be attracted away from the centre of the anion vacancy and towards that positive ion situated along the tetragonal axis of the defect. Thus, although we expect a hyperfine interaction with magnesium ions in any of the five nearest-neighbour cation sites it is likely to be strongest with that nucleus along the tetragonal axis. Thus there are four equivalent nuclei and one non-equivalent nucleus. The expected contribution of hyperfine lines from the single cation is about 10% of the total since this is the probability of the P^- centres having that particular site occupied by an Mg nucleus. There should also be a contribution amounting to 32% of the total intensity arising from those centres in which one or more of the four equivalent cation sites are occupied by ^{25}Mg nuclei. The remainder of the P^- centres have no ^{25}Mg nuclei in the five nearest-neighbour Mg^{2+} sites, and the magnetic resonance spectrum will consist of a single line. These expectations are substantially supported by the experimental results at both X and Q bands. An interesting feature is that with H_0 along [100] the $g_\parallel$ line at Q band is broadened appreciably compared with the two coincident lines from those P^- centres perpendicular to H_0. The [111] crystal axis is equally inclined to each of the [100] axes and constitutes the most appropriate axis along which to investigate the hyperfine interaction. In this orientation six weak lines with separations of 17·5 G are centred on the strong principal line (To, Stoneham and Henderson (1969)). It is significant that this hyperfine splitting is almost four times as large as that of the F^+ centre and also that the intensity of these lines is approximately 10% of the total intensity in the spectrum. In addition, the line shape of the central line on an expanded scale reveals structure which also seems likely to be associated with the hyperfine interaction with ^{25}Mg. This constitutes strong evidence that the five nearest-neighbour cation sites are inequivalent in the case of the P^- centre. Formally, therefore, we identify the large hyperfine splitting with those P^- centres which have one $^{25}Mg^{2+}$ ion along the defect axis. The increased hyperfine constant indicates the extent to which the electron is localized upon this ion relative to the other four. Since the electron is displaced relative to the centre of the anion vacancy, it would appear that the hyperfine interaction with the four equivalent nuclei should be appreciably more anisotropic than in the case of the F^+ centre. Since it is also a small interaction, the lines appear as shoulders on the central resonance line.

The separation between successive hyperfine lines may be written as:

$$\Delta H = a + b(3\cos^2\theta - 1),$$

where a and b are the isotropic and anisotropic components of the hyperfine interaction in the spin Hamiltonian, eqn. (11), with $S = \frac{1}{2}$, and $I = \frac{5}{2}$. Thus with magnetic field parallel to [111] we see that our measurement gives directly $a_1 = 17.5$ G. Measurements along [100] and also [110] give values for $b_1 = 0.7$ G. It is more difficult to estimate the corresponding values of the second hyperfine interaction, although one easily recognizes that this weak interaction is less than that observed on the F^+ centre. The g values quoted in table 17 and the hyperfine structure discussed above emphasize how similar are the P^- centres to the surface F^+ centres (S centres) discussed in detail by Nelson et al. (1967). This is not surprising since one is again dealing with an axially symmetric site and only five nearest-neighbour cation sites. The similarity, at least as far as the g values are concerned, is very pronounced. No hyperfine structure has yet been reported for the P^- centres or the S centres in calcium or strontium oxide.

The optical properties of these defects have been obscured by the many other optical bands present in annealed crystals. Theoretical studies by To et al. (1969) were instrumental in indicating the nature of the bound states of P^- centres. They calculated the binding energy of the electron in the dipole field of two vacancies with charges $-z|e|$ and $+z|e|$ respectively, separated by a Å in a medium with dielectric constant, ε. The dipole moment of the pair is of magnitude $D = 2.53 \times 10^{18} R$, e.s.u.-cm, where R is the 'effective length' of the dipole in the dielectric medium and is equal to $za/0.529\varepsilon$ Å. The Schrödinger equation for the electron bound to this dipole is:

$$\left[-\frac{h^2}{2m}\nabla^2 + \frac{z|e|}{\varepsilon}\left(\frac{1}{|\mathbf{r}-\mathbf{R}_+|} - \frac{1}{|\mathbf{r}-\mathbf{R}_-|}\right)\right]\psi = E\psi. \quad (51)$$

This is an 'effective mass' model which should be good for weakly bound centres. A strict effective mass theory uses ε_0 as the dielectric constant; however, F^+-centre models use ε_∞. The eigenvalues of eqn. (1) have been tabulated as a function of R by Wallis, Herman and Milnes (1960) and the required energy values are obtained from their tables by:

$$E_B = \text{tabulated } E_b \times \frac{R_\infty^2}{a}.$$

This implies that as $a \to \infty$ at constant R, the binding energy $E_B \to 0$. The appropriate values of E_B obtained by To et al. (1969) are given for the alkaline earth oxides in table 17. In addition, we note that the 'effective mass' approximation should be good since the value of E_B/E_{gap} indicates weak binding. No other bound states exist within 0.01 eV of the bottom of the conduction band. Although these calculations are only approximate, they do give important information about the optical absorption spectrum to be expected from these defects. In principle the defects should in all cases be bleachable with photons of appropriate energy. The results of King and Henderson (1967) substantially support this view in magnesium oxide. These latter workers generated the P^- centres by annealing samples at 720°C for 4 hours followed by $\frac{1}{2}$ hour at 900°C. The samples were then X-irradiated for 1 hour with 33 kV X-rays from a Cu target. This treatment optimized the concentration of P^- centres and reduced the F^+ concentration to undetectably small limits. The P^- centre concentration was then monitored continuously whilst bleaching with monochromatic light in the wave length range 2500–6500 Å. The intensity of P^- centre line was then measured relative to one of the $\frac{1}{2} \leftrightarrow -\frac{1}{2}$ transitions of the Mn^{2+} ion since this spectrum is insensitive to the bleaching process. The results indicate that the peak bleaching efficiency occurs at about 3·6 eV. This energy is rather larger than published predictions of To *et al.* (1969) but suggests the essential correctness of the model. Preliminary optical reflectance spectra from powders indicate that the electron in the surface F^+ centre is in a trap of at least 3·0 eV in depth (Tench and Nelson 1967 b). No similar studies have yet been made on the other oxides.

There are many anisotropic ESR spectra to be observed in neutron irradiated CaO. Hall (1975 b) has observed several defects with $S = 1$ and rather low symmetry: on annealing these spectra are either destroyed or revert to the F^+ pair spectrum reported by Tanimoto *et al.* (1965). In addition there are several spectra due to trapped hole centres involving rather complex vacancy aggregates (Hall 1975 a). A particularly interesting spectrum has $s = \frac{1}{2}$ with principal g tensor axes, $X = [0\bar{1}\bar{1}]$, $Y \sim [\bar{1}11]$ and Z tilted 6·5° about [212] towards [111] (Bessent 1969). This spectrum clearly is due to an intrinsic centre which Bessent suggests has the F_4 configuration.

More complex F-like centres resulting from annealing magnesium oxide crystals irradiated to high doses have also been observed. One of these has one axis along a body diagonal, and the magnetic resonance spectrum indicates an electron trapped at a body-centred anion vacancy to be involved (Wertz *et al.* 1961). The spectrum of this defect fits the spin Hamiltonian of eqn. (11), where the defect axis is now the [111] axis of the crystal, $g = 2 \cdot 0011$, $A = 42$ G, $B = 1 \cdot 5$ G, $S = \frac{1}{2}$ and $I = \frac{5}{2}$. The hyperfine structure seemingly comes from the 10% abundant ^{25}Mg nuclei. The proposed model implies that as a result of the very heavy neutron irradiation, regions of body-centred symmetry rather than face-centred symmetry are formed. Although such a structure seems improbable, no more satisfactory alternative model has been forthcoming.

There are many complex changes in the optical spectrum of magnesium oxide which occur in parallel with the production of the F^+ centre. A typical set of spectra are shown in fig. 33, where it is apparent that many new bands grow at positions similar to those observed in very heavily irradiated crystals. There is now evidence which very strongly supports the suggestion that these bands are due to F-aggregate centres similar to the F_2, F_3 and F_4 centres in the alkali halides. This evidence has largely been obtained from optical measurements at low temperatures where sharp structure is observed in association with the broad bands. The underlying theory necessary to understand the presence of this structure at low temperatures and the way in which the sharpest lines may be used as a sensitive probe into the structure of defects has been outlined in § 2.

6.4. *Zero-phonon transitions in magnesium oxide*

Zero-phonon and phonon-assisted transitions attendant upon broad colour centre bands in magnesium oxide were first observed by Wertz *et al.* (1963). The fine structure was present in as-irradiated crystals and more abundantly in crystals annealed subsequent to irradiation. The obvious similarities between these spectra and those reported in the alkali halides prompted the suggestion that the coloration results from the presence of *F*-aggregate defects (Wertz *et al.* 1964). There have since been a large number of observations and an approximate understanding of the defect structures involved has been achieved. In order that these lines may be used to probe the lattice vibrational processes and the defect structure they must of necessity be extremely narrow. Not surprisingly, the lines observed in neutron irradiated magnesium oxide are generally broader, sometimes by a factor of ten, than the zero-phonon lines in the alkali halides. Suitable control of the width of particular zero-phonon lines may be obtained by the correct choice of conditions of irradiation dose and of annealing temperature. The most general treatments resulting in small half-widths (20 cm^{-1}) are: (*a*) irradiation to doses of order 10^{19} nvt at a temperature of $\sim 600°$C or (*b*) irradiation to doses $< 10^{19}$ nvt, followed by annealing at 600°C. Some typical spectra are shown in fig. 33.

Fig. 33

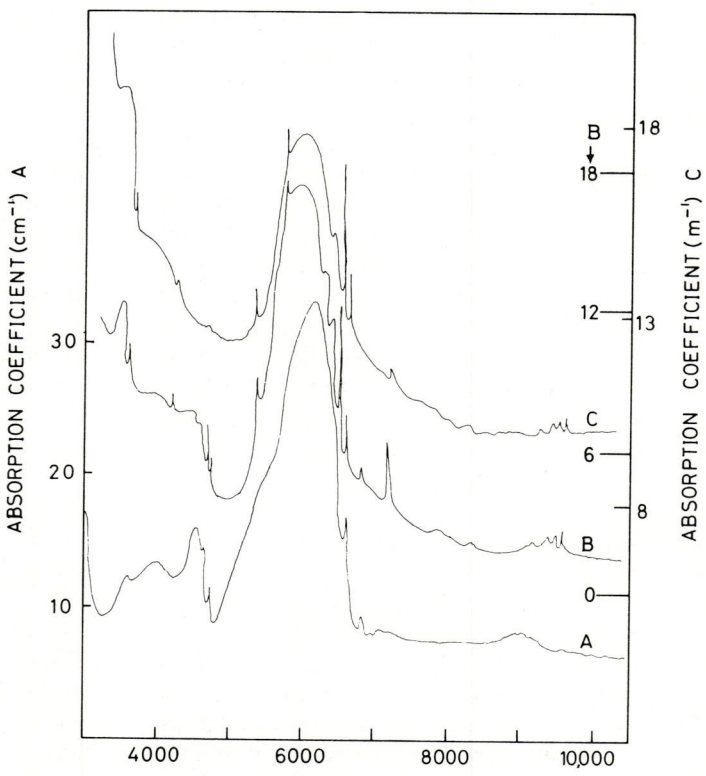

Illustrating the fine structure observed at low temperature ($<77°$K) on colour-centre absorption bands in MgO. (*A*) annealed at 400°C, subsequent to fast neutron irradiation at 150°C to a dose of $6·8 \times 10^{19}$ nvt > 1 Mev; (**B**) neutron irradiated at 600°C to a dose of $6·8 \times 10^{19}$ nvt > 1 Mev; (C) annealed at 600°C subsequent to neutron irradiation at 45°C to a dose of 7×10^{18} nvt > 1 Mev.

The half-widths of all the zero-phonon lines decrease both with increasing irradiation temperature and increasing annealing temperature. For a given dose the lines are narrower in crystals irradiated at elevated temperatures than in crystals irradiated at 150°C (or lower temperatures) and subsequently annealed at the same elevated temperature. In addition the half-widths increase with increasing dose for a particular irradiation temperature. Detailed investigations of the linewidths (King and Henderson 1966, King 1967) suggest that although aggregate defects exist in crystals heavily irradiated below 150°C the zero-phonon lines are so broadened as to be observable. Optical zero-phonon lines are inhomogeneously broadened by the presence of dislocations and point defects (Stoneham 1967, Hughes 1968).

Thus we suggest that the linewidth variations noted above are due to random strains resulting from isolated interstitial ions and small clusters of interstitial ions. The strain in annealed crystals is reduced via interstitial/vacancy recombination and interstitial aggregation, a view supported by the recent electron microscope and other studies by Bowen and his associates (Bowen and Clarke 1964, Henderson and Bowen 1971, Morgan and Bowen 1967). In crystals irradiated at 150°C, the interstitials are present as point defects and also in small clusters of approximately 50 Å diameter. The clustered interstitials are little affected by annealing below 800°C, although interstitial/vacancy recombination does reduce the lattice strain as demonstrated by X-ray measurements. In magnesium oxide irradiated at 600°C the interstitial dislocation loops are usually about 1000 Å in diameter: these correspond to loop sizes in crystals irradiated at 150°C and annealed at 1100–1200°C. In addition, the irradiation-induced macroscopic and X-ray growth is very small. These effects are consistent with a reduced level of macroscopic strain in the crystals and consequently narrower zero-phonon lines. Furthermore, the observations of narrower linewidths in crystals irradiated at 600°C indicate that diffusion processes are enhanced by irradiation at high temperatures. Such enhanced diffusion processes also reduce the defect concentration as indicated by the reduced intensity in the zero-phonon line (except the 6490 Å line) after irradiation at higher temperature. Chen and Sibley (1969) have recently investigated this annealing behaviour in greater detail: apparently simple kinetic behaviour is not observed, and thus it is difficult to make any confident conclusions from such studies about the nature of the defects involved. Broadening of the zero-phonon lines due to the presence of the different magnesium isotopes will also contribute to the observed linewidth, although the extent of such an effect has not been measured.

6.3.1. *Vibrational structure*

The preceding qualitative discussion of the electron phonon interaction (§ 2.2) takes no account of the particular type of mode involved. Three kinds of normal mode exist in a crystal containing defects: (*a*) normal lattice modes closely similar to those of the perfect crystals and which are only modified close to the defect, (*b*) resonant or pseudo-localized modes having frequencies within the continuum of lattice modes and maximum amplitude near the defect. Even a small density of such modes can give appreciable contributions to the vibrational structure observed, (*c*) localized or non-resonant modes having frequencies outside the continuum and amplitudes which decrease exponentially with distance from the defect. In general, all three types of mode may exist and contribute to the vibrational sidebands of a particular defect. The vibrational energies interacting with the defect electronic states may be compared with the energies of the lattice phonons with high density of states at the

Brillouin zone boundary, as derived from phonon dispersion curves. Such curves have been determined for magnesium oxide by Peckham (1967), using the slow-neutron-scattering method.

The assignment of the vibrational modes for all the zero-phonon lines in magnesium oxide are shown in table 18. In some instances zero-phonon lines are sufficiently close together to hamper the unambiguous assignment of the vibrational

Table 18. Phonon-assisted transitions and zero-phonon lines in MgO

Zero-phonon line		Phonon-assisted line (cm^{-1})	Energy difference (cm^{-1})	Phonon Assignment	References
Å	cm^{-1}				
3618	27651	27839	188	Local mode A	Ludlow and Runciman (1965)
		27943	292	TA [100], TA [110] B	
		28014	363	2 × A	
		28052	401	TO [110] or TO [000]	
		28114	463	LA [110]	
		28227	576	3 × A or 2 × B	
		28285	634	TO [111] + (TA [100] or TA [110])	
4685	21345	21594	249	Local mode	King and Henderson (1966), King 1967
		21758	413	LO [110], TO [100]	
		22183	838	2 × LO [110] or 2 × TO [100]	
4700	21276	21701	425	LO [110], TO [100]	King and Henderson (1966), King (1967)
6419	15579	15853	274	TA [111]	King (1967)
		15936	357	TO [111]	
		16000	421	LA or TO [100], LO [110]	
		16041	462	LA [110]	
		16279	700	2 × TO [111]	
		16447	868	2 × LA [100]	
6491	15406	15691	285	TA [111]	King (1967)
		15753	347	TO [111]	
6706	14912	15204	292	TA [110]	King and Henderson (1967)
7058	14168	14384	216	Local mode	King (1967)
		14453	285	TA [110]	
		14522	354	TO [111]	
		14556	383	TO [110]	
9931	10070	10450	380	TO [110]	Ludlow (1966)
		10840	770	2 × TO [110]	
		11050	980	2 × TO [111]	
		11210	1140	3 × TO [110]	
		11550	1480	3 × LO [100]	
		12030	1960	4 × TO [110]	
10450	9570	9850	280	TA [111] or TA [110]	Ludlow (1966)
		9950	380	TO [110]	
		10210	640	TA [111] + TO [110]	
		10330	760	2 × TO [110]	
		10690	1120	3 × [TO 110]	

structure. This is particularly true for three pair of lines at 4685 Å and 4700 Å, 6420 Å and 6490 Å and at 9930 Å and 10 450 Å, although it has been possible to resolve most of the uncertainties. Figure 34 shows the conditions under which a more detailed resolution of the phonon structure of the 6420 Å and 6490 Å line is

Fig. 34

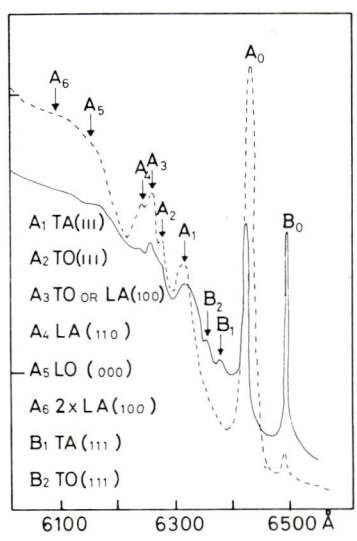

Analysis of the phonon sidebands associated with the zero-phonon lines at 6420 Å and 6490 Å in magnesium oxide crystals subjected to two different treatments. (After King 1967.)

possible (King 1967, King and Henderson 1966). The system of lines near 10 000 Å has been analysed by Ludlow (1966) and by others (Wertz et al. 1964, Stettler et al. 1965, Kazumata et al. 1965). Examination of the vibrational structure for several systems suggests that the electronic states may be coupled to pseudo-localized modes at the defect. In general, however, the vibronic coupling is to several phonon modes and propagation directions rather than to a single dominant phonon mode. The transition processes may be expected to be coupled to a variety of different modes when the relatively complex aggregate defects are involved (Pierce 1964). Thus these spectra imply that the defects are complexes of similar form to the F-aggregate centres in the alkali halides for which vibrational sidebands have also been observed (Hughes 1966).

6.4. Symmetry of colour centres in magnesium oxide

Several investigations of uniaxial stress splitting of zero-phonon lines have now been reported, and three different symmetry systems are now known to exist. These are (using Kaplianskii's notation): orthorhombic I centres with electric dipole moment along a [110] symmetry axis, trigonal centres with a symmetry axis along a [111] direction and electronic degeneracy associated with an $A \leftrightarrow E$ transition, and monoclinic I centres with an electric dipole oscillator along the [112] direction. One should also expect cubic and non-cubic centres associated with impurities; and $3d^3$ ions V^{3+}, Cr^{3+} and Mn^{4+} in cubic and distorted symmetries have been subjected to detailed investigations as noted in § 3.

6.4.1. Orthorhombic centres

The splitting patterns of the zero-phonon lines at 3618 Å, 4132 Å, 4685 Å, 4700 Å, 6706 Å and 10 450 Å are consistent with centres having orthorhombic symmetry with linear electric dipole oscillators along the [110] axis. The theoretical splittings, intensities and polarizations for such centres are given in fig. 35 where we use as an

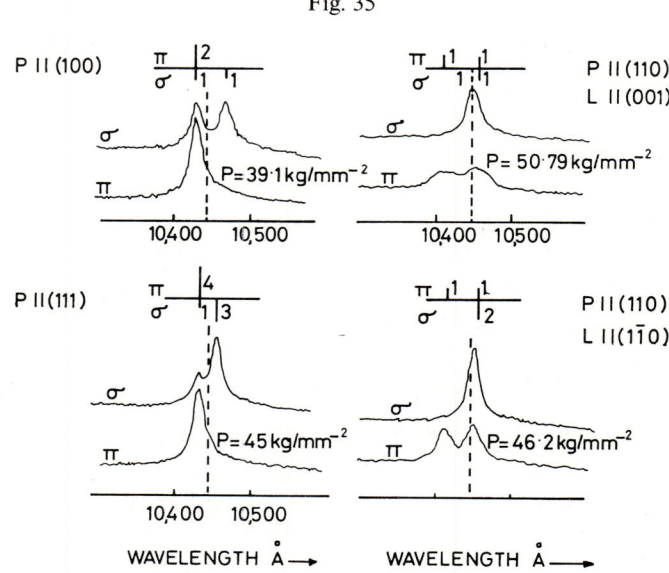

Fig. 35

Stress splitting patterns for the 10 450 Å line in neutron-irradiated magnesium oxide. (After Ludlow 1966.) (By courtesy of the Institute of Physics and the Physical Society.)

example Ludlow's results for the 10 450 Å line; since a cubic crystal becomes biaxial under [110] stress, two patterns are shown corresponding to [001] and [110] directions of propagation of the viewing light.

We see that the number and the intensities of the components agree with the theoretical expectations for a centre with orientational degeneracy associated with orthorhombic symmetry (D_{2h}, D_2 or C_{2v}) and an electric dipole transition along the [110] direction. Similar results were obtained for the 10 450 Å line by Stettler *et al.* (1966).

Table 19. The stress parameters A_1, A_2 and A_3 (in units of $cm^{-1} kg^{-1} mm^2$) and strain parameters B_1, B_2 and B_3 (in units of $cm^{-1}/1\%$ strain), for orthorhombic centres in magnesium oxide.

Line position Å	A_1	A_2	A_3	B_1	B_2	B_3	References
3618	0·98	0·70	0·38	42·1	29·7	42·0	Ludlow and Runciman (1966)
4132	0·62	0·55	0·55	38·1	20·4	8·9	Ludlow and Runciman (1966)
4685	−0·17	0·54	0·40	25·3	12·4	6·4	King and Henderson (1966)
4700	−0·15	0·52	0·49	26·2	10·4	8·2	King and Henderson (1966)
6706	+0·11	0·05	0·45	22·1	4·45	4·1	King (1967)
10450	−0·6	0·38	0·33	15·0	4·4	−11·5	Ludlow (1966)

A check on the symmetry assignment may be made by examining the energy shift of each component in the patterns. The change in the potential at the defect can be represented by an operator V which may be expressed in terms of macroscopic stress or strain components as:

$$V = \sum_{ij} A_{ij}\sigma_{ij} = \sum_{kl} B_{kl}\varepsilon_{kl}, \tag{52}$$

where σ_{ij} and ε_{ij} are the stress and strain tensors. A_{ij} and B_{kl} are electronic operators which are linearly related by the elastic constants of the crystal. For orthorhombic centres only the three stress parameters A_1, A_2 and A_3 are required. Equivalent relationships in terms of the independent strains along the principal axes of the centre (not the crystal) are given by:

$$\begin{aligned} B_1 &= c_{12}A_1 + (c_{11} + c_{12})A_2 + c_{44}A_3 \quad \text{along } [110], \\ B_2 &= c_{12}A_1 + (c_{11} + c_{12})A_2 - c_{44}A_3 \quad \text{along } [1\overline{1}0], \\ B_3 &= c_{11}A_1 + 2c_{12}A_2 \quad \text{along } [001], \end{aligned} \tag{53}$$

where C_{11}, C_{12} and C_{44} are the elastic constants for magnesium oxide. Comparison of the data on six orthorhombic centres is given in table 18. The constants which parametrize the stress spectra of the 4685 Å and 4700 Å lines are almost identical, suggesting that the defects involved are very similar. The relative intensities of these lines do not change in parallel fashion for any variation in heat treatment. Consequently, the lines are not due to two electronic transitions within the same defect. It seems more probable that one line is associated with an intrinsic lattice defect, and the other with the same defect perturbed by an impurity atom along one of the defect symmetry axes. Since the strain component B_2 along the [1$\overline{1}$0] direction changes most the impurity atom may be situated along this axis rather than along the [110] direction of the electric dipole oscillator. A cation vacancy in the same position would not alter the symmetry, although it might be expected to produce a larger shift in B_2 than has been measured.

Ludlow (1966) suggested that the 10 450 Å and 3618 Å lines are due to transitions within the same centre. Detailed results show that although the experimental scatter is quite large the intensities change in parallel (Henderson, unpublished). This behaviour is in agreement with Ludlow's suggestion, although it is hardly conclusive. Kazumata et al. (1965) studied the temperature dependence of the half-widths, $H(T)$, of these zero-phonon lines. Their results obey a coth $hv_g/2kT$ relation with different values of v_g for the two lines. They deduced from this that the lines were *not* from the same centre. This is not justified since the half-widths of the lines depend on the strength of the phonon-defect interaction and on the effective density of phonon states, neither of which need be the same for different transitions within the same centre. King (1967) investigated the dose dependence of these two lines and observed that the maximum intensity of the lines occurred at a lower dose than the maximum in the F^+-centre concentration. He argued that this is suggestive of defects involving impurities. However, Ludlow (1966) had previously investigated impurity effects and concluded, in the absence of any dependence of the lines on impurities, that they were due to intrinsic lattice defects. The defects were supposedly produced by the displacement of either or both magnesium and oxygen ions in the lattice. Preliminary results for the Zeeman effect on the 3618 Å show that no detectable splittings occur up to 150 Kg. The absence of any such splitting can be explained by the centre having an even number of electrons (diamagnetic), or an odd number of electrons with the selection rule that only $\Delta M_s = 0$ transitions are allowed (Ludlow 1966).

Unfortunately the piezo-spectroscopic technique does not allow unequivocal identification of the defects responsible for the observed transitions. A speculative set of structures is proposed in fig. 36. Detailed assignments of the centres will require considerable help from either the Zeeman effect, bleaching experiments or a correlation of electron spin resonance absorption and optical absorption. So far only two orthorhombic centres have been detected using paramagnetic resonance (Wertz et al. 1961, Henderson and King 1967), and neither can be correlated with the zero-phonon lines discussed here. If $S = \frac{1}{2}$ for any of the defects, we might expect that the resonance absorption would be easily detected. However the defects involved are present in small concentrations ($\sim 10^{13}$–10^{14} cm^{-3}), implying that resonance absorption from such defects would not be detected under the very strong F^+ centre spectrum. It seems reasonable to base the models on the alkali halides F_2-like centres with varying electron populations. For electrostatic neutrality four electrons are required. Since the defects which give rise to the 10 450 Å and 3618 Å lines are produced during irradiation at a rate close to 5% of the F-centre rate, we ascribe them as primary radiation products. (It will be remembered that the silicon divacancy is produced at a similar rate compared with the single vacancy, Corbett 1966.) We speculate that the transitions are from the ground state of an F_2-like centre containing four electrons (fig. 36). The two optical transitions will then occur from a quantum state with $n = 2$. A particle-in-a-box model yields a ratio for the transition energies for $\Delta n = 2$ and $\Delta n = 1$ quantum states as 12/5, which is reasonably close to the observed ratio.

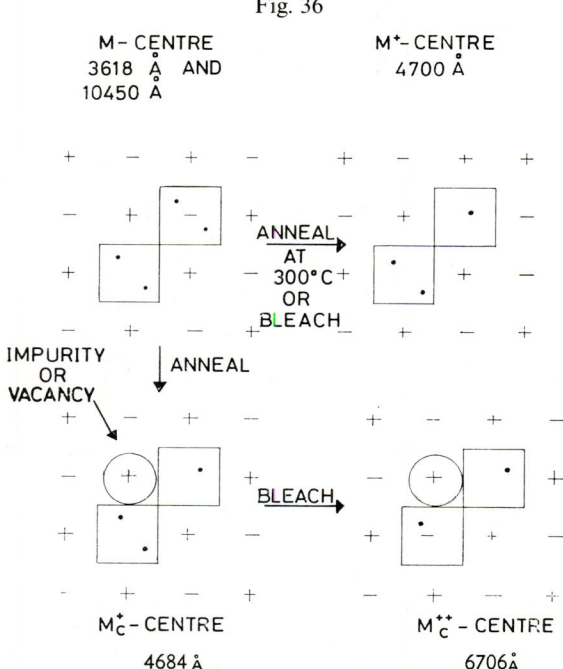

Fig. 36

Some models for F_2-like centres in the alkaline earth oxides, based upon observations in MgO.

Chen and Sibley (1969) have observed that the 4685 Å and 4700 Å lines increase in intensity at temperatures ($>300°C$) such that the 3618/10,450 Å pair are decreasing in intensity. A defect formed from the F_2 centre by thermal ionization or by a further stage in the aggregation would account for this behaviour. Thermal ionization would produce a paramagnetic defect, but the optical absorption intensities indicate such a small concentration that it would be undetectable. Consequently, in fig. 36 we propose this to be the model associated with one or other of the 4687 Å and 4700 Å lines. The model proposed for the 6706 Å line is then based upon the optical bleaching behaviour reported by King (1967). The results of such behaviour are shown in fig. 37; a simple electron transfer apparently accounts for most of the defects as shown in fig. 36.

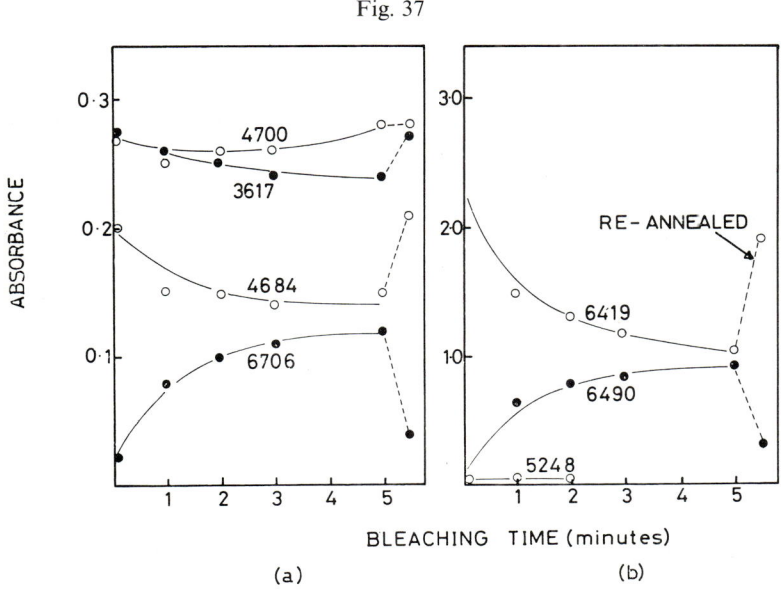

Fig. 37

Optical bleaching of colour-centre zero-phonon lines in MgO. (a) orthorhombic centres, (b) trigonal centres.

One criticism of the above models is that little use has been made of the known impurities to account for observations. At the present time it does seem clear that the lines are associated mainly with intrinsic defects. Even in the purest crystals, where transition metal impurities are very low, the defects may be produced in comparable concentrations to those observed in the earlier highly impure crystals. This would seem to justify the present conclusion as to the involvement of intrinsic rather than extrinsic defects.

6.4.2. *Trigonal centres*

The lines at 5248 Å, 6420 Å and 6490 Å are associated with trigonal centres having $\langle 111 \rangle$ orientations and electronic degeneracy associated with $A \leftrightarrow E$ electric dipole transitions (King and Henderson 1966). The intensities of the three lines associated with the trigonal centres vary in an apparently unconnected way during heat treatment. They cannot, therefore, be due to transitions within the same defect, and we seek three different centres in order to account for the results. Typical half-widths for these lines of order ~ 40 cm^{-1} resulted in difficulties in resolving all the components of the stress spectra; these were overcome only by applying stresses up to 70 Kg mm^{-2}. The observed stress spectra for the 5248 Å line are in good agreement with the theoretical splittings and polarizations for an $A \leftrightarrow E$ transition in a centre of trigonal symmetry, despite incomplete resolution of some of the components (King 1967). Because of the smaller half-widths, the components of the 6420 Å and 6490 Å lines are better resolved in the stress spectra. For trigonal centres with $A \leftrightarrow E$ transitions, the energy shifts of the components may be expressed in terms of four independent parameters A_1, A_2, B and C. The first two describe the removal of orientational degeneracy and the last two the removal of electronic degeneracy. These stress parameters may be redefined in terms of the independent strains along the principal axis of the defect using eqn. (52). The strain parameters for these three lines are compared in table 20.

Table 20. The stress and strain parameters for trigonal centres in magnesium oxide

Parameters		5248 Å†	6240 Å‡	6490 Å‡
Stress parameters (cm^{-1} Kg^{-1} mm^2)	A_1	0·18	0·077	0·055
	A_2	−0·18	−0·190	−0·165
	B	−0·08	−0·130	+0·041
	C	−0·13	−0·190	−0·106
Strain parameters cm^{-1}/10^{-3} strain	$(C_{11} + 2C_{12})A_1 + 2C_{44}A_2$	3·3	−2·3	−2·6
	$(C_{11} + 2C_{12})A_1 - C_1 A_2$	11·3	6·7	5·3
	$(C_{11} - C_{12})B + C_{44}C$	−3·7	−5·9	−0·8
	$\sqrt{2}(C_{11} - C_{12})B \frac{1}{\sqrt{2}} C_{44}C$	−0·97	−1·8	+2·4

† After King (1967); less accurate results than Ludlow's were reported for the 6420 Å and 6490 Å.
‡ After Ludlow (1968).

As indicated in fig. 38 the stress spectra of the 5248 Å line show a marked stress-induced dichroism at 4°K, i.e. the total π and total σ intensities are unequal. This is not expected for simple anisotropic centres since reorientation of the centres is not possible at low temperatures. The dichroism may be explained if the electronic ground state is the orbitally degenerate E state whilst the excited state is the A state. This orbital degeneracy will be removed by the stress in all cases except for those centres whose trigonal axis coincides with the stress direction. If the E state splitting is Δ then at a temperature T a population ratio $\exp(-\Delta/kT)$ will be established between the two component levels; the intensity of the line associated with transitions from the lower level will be enhanced at the expense of that from the upper. At 4°K, $kT \sim 3$ cm^{-1}, so that a very small value of Δ could drastically deplete the upper level, resulting in only one level being occupied. Since Δ increases with increasing stress, the dichroism will saturate at high stresses. In the present case measurements at 77°K show that splittings of 15 cm^{-1} can be achieved, and the population ratio between the two levels will be $\sim 1/300$ at 4°K. The observed dichroism in the 5248 Å line is consistent with the trigonal centre having optical dipole moments along the [110] and [112] directions in the plane perpendicular to the trigonal axis. As shown in fig. 38 the absence of one component at 4°K demonstrates that the [112] dipole is the lower in energy, since it will give rise to absorption in both π and σ polarizations.

Fig. 38

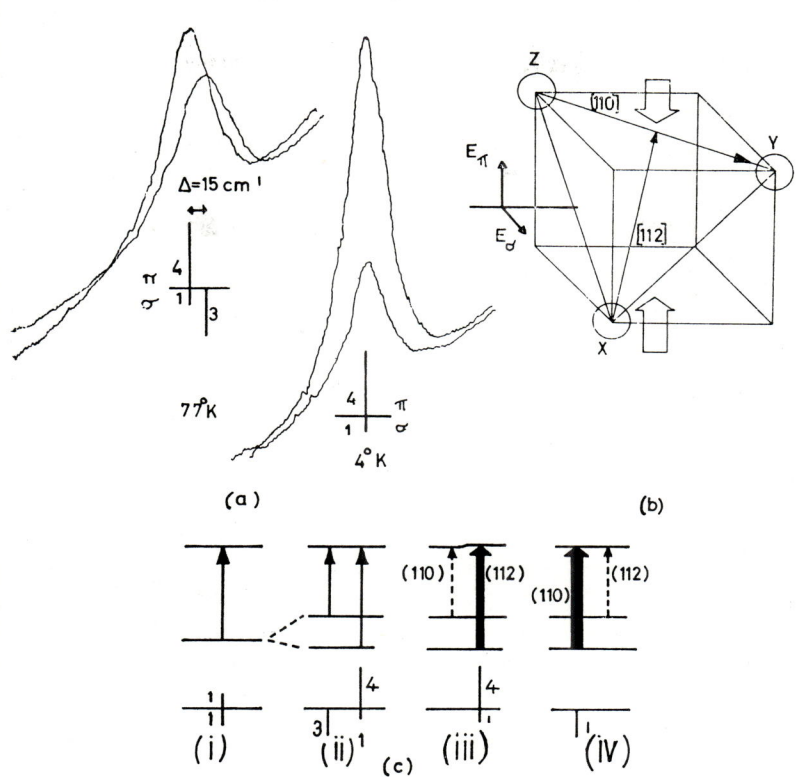

Showing (a) the stress-induced dichroism of the 5248 Å line under a [100] stress of 52 Kg/mm^2 and (b) one of the four equivalent orientations of a trigonal centre under a [100] stress. In (c) the theoretical expectations for an $^2E \rightarrow {}^2A$ transition at 77°K under (i) zero stress and (ii) high stress, are illustrated. At 4°K (iii) obtains if the $|E_s\rangle$ state is lower in energy than the $|E_y\rangle$ state ([112] dipoles), whereas (iv) is expected when the $|E_y\rangle$ state is lower ([10$\bar{1}$] dipoles).

The [110] dipole absorbs only for σ polarizations. Thus we conclude that the defect has an E ground state. We note that this behaviour is very similar to that observed for the F_3 centre in KCl (Silsbee 1965), and LiF (Hughes and Runciman 1965).

The lines at 6420 Å and 6490 Å are not dichroic under stress (Henderson and King 1967, Ludlow 1968) even though in the case of the 6420 Å line the splittings are large enough ($\sim$37 cm^{-1}) for such an effect to have been detected at 77°K. Thus it is apparent that they are associated with defects in which the A state is the ground state. High resolution measurements (Ludlow 1968) on the 6490 Å line show that the line is in fact a doublet in the absence of stress, with a separation between the components of 9 cm^{-1}. No dichroism is observed at 4°K in the absence of stress; clearly the splitting occurs in the orbitally-degenerate excited state.

Ludlow (1968) originally attributed this splitting to spin–orbit coupling in the excited E state, a view consistent with the stress dependent intensities of the components. Subsequent work showed this to be incorrect, in fact in some crystals the doublet character is not observed. Instead a single line is observed, the stress splitting pattern being in accord with the reports of King and Henderson (1966) and King (1968). Ludlow (1970) suggests that the splitting is due to a nearby charged defect. Preliminary MCD studies at low temperature confirm that any spin–orbit structure in the excited state must be very small (Merle d'Aubigné 1969, private communication).

The interpretation of the stress spectra for the three lines at 5248 Å, 6420 Å and 6490 Å involving $A \leftrightarrow E$ transitions within centres of trigonal symmetry suggests centres similar to the alkali halide F_3 centre. Henderson and King (1967) and Ludlow (1968) have suggested models (fig. 39) based upon three anion vacancies on the (111) plane and trapping between five to seven electrons to account for these three lines.

These defect structures are referred to for identification purposes as F_3^+, F_3 and F_3^- centres. Unlike the P^- centre, which has no bound excited states and gives a broad ionization spectrum, the F_3^- centres must have bound excited states for sharp line spectra to be observed. It is not unreasonable that such bound states should exist since the extra electron is shared between a greater number of vacancies for these larger defects and hence will be more strongly bound that in the P^- centre. An alternative set of models first suggested by Ludlow (1968) involving the cation vacancy along the trigonal axis of the defect would require between three and five electrons. All the models have C_{3v} symmetry and the only orbital states are those described by the irreducible representations A_1, A_2 and E of that group. The stress results do not differentiate between A_1 and A_2 states although other considerations may do so. Ludlow (1968) suggests that in the presence of the cation vacancy an effective repulsion experienced by electrons near the centre of the defect may result in the E orbital being lower than the A orbital. This is not considered by the present authors.

Fig. 39

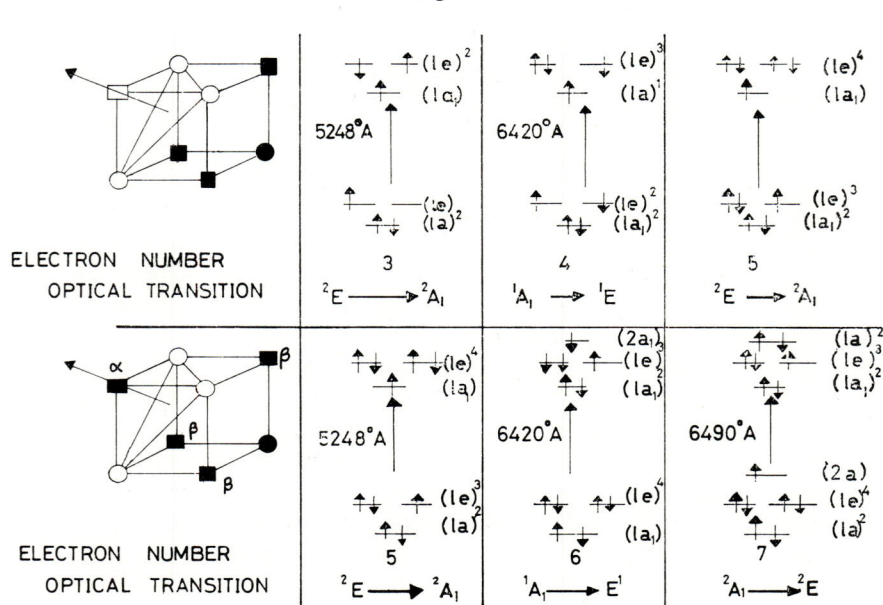

Some models and expected transitions for the F_3-like centres in the alkaline earth oxides.

The electronic states of the defects may be constructed using a molecular orbital or valence band approach. The former technique handles the excited states the better and has been used extensively to calculate the one electron orbitals of the numerous F_3-like centres in the alkali halides (Van Doorn 1962, Hughes 1966). We see from fig. 39 that similar considerations yield three defect models with ^2E ground states characteristic of the defect associated with the 5248 Å line; there is no *a priori* reason for selecting any one of the three models. It is, however, obvious that the defect species must be paramagnetic since they contain odd numbers of electrons. This is confirmed for the 5248 Å line by the recent studies of temperature and magnetic field variations of the optical Faraday rotation spectrum of this line (Kemp et al. 1967 a). The rotation pattern is opposite in sign to that corresponding to a one electron centre with negative spin orbit coupling constant. Thus Glaze and Kemp (1969), suggests that a hole in the (1e) level is involved, and apparently the required centre is either the F_{3c}^+ centre or F_3^+ centre. To differentiate between the models one would need to investigate the hyperfine structure in a paramagnetic resonance spectrum. As in the ^2E-ground state resonance of the R centre in KCl (Krupka and Silsbee 1966) one expects strong electron–phonon coupling and Jahn–Teller distortions to result in very short relaxation times. Thus the resonance spectrum may not be observed even at the lowest temperatures, and it is not surprising that no such spectrum has yet been reported. However, one may differentiate between the defects on the basis of their likely excited states. For the F_{3c}^+ centre the possible states are $^2A_1$ and $^2A_2$, whilst for the F_3^+ centre it is $^2A_1$.

Referring to fig. 38 we note that a distortion of the equilateral triangle caused by an increase in the angle x, lowers the $|E_x\rangle$ state relative to the $|E_y\rangle$ state. This is the situation under [001] stress, and the polarization of the splitting components in this case shows that the transition from the lower state has an optical dipole moment along the [112] direction. Such a moment comes from the $|E_y\rangle$ state if the excited state is $^2A_1$, or the E_x state if it is $^2A_2$, hence this result favours $^2A_2$ as the excited state. Furthermore the oscillator strength of the $^2E_2 \rightarrow {}^2A_2$ transition is very much greater than that of $^2E \rightarrow {}^2A_1$ (Silsbee 1965). Thus the results described above favour the F_{3c}^+ centre as the defect responsible for the 5248 Å line.

Simplified calculations of the energy of the $E \rightarrow A$ transition of the F_3 centre in the alkali halides have been used to elucidate the main features of the molecular orbital approach (Hughes 1966). However, the F_{3c}^+ centre discussed above is crudely analogous to the H_3 molecule embedded in a dielectric continuum. The results of Hirschfelder's calculation for the H_3 molecule may be scaled in the continuum approximation by dividing energies by ε^2 and multiplying distances by ε. The energy of the $^2E \rightarrow {}^2A_2$ transition scaled in this way is about 2·8 eV, which is reasonably close to the energy of the observed line.

One interesting feature of this line is that no phonon-assisted structure is observed; furthermore, no evidence of the Jahn–Teller effect so apparent in the F_3 centre in the alkali halides has yet been reported.

The 6420 Å line has the A state as ground state and is apparently diamagnetic (Kemp, unpublished). Consequently it would appear to be associated with a defect containing an even number of electrons. The most favourable models are the F_{3c} centre and the F_3 centre shown in fig. 39. Bearing in mind that the $(a_1)^2(e)$ ground state configuration for the F_c^+ centre predicts the ^2E level consistent with the stress results, it seems natural that the centre with four electrons has the configuration $(a_1)^2(e)^2$. This configuration is split by electron–electron interaction into the three states $^1A_1$, ^1E and $^3A_2$. According to Hund's rules for the states of a single configuration, the $^3A_2$ state should lie lowest, since this state has the greatest multiplicity. No spin resonance spectrum attributable to this centre has been observed. Such a centre should be easily observed even for a small concentration of centres, since the $S = 1$ state will be split into the magnetic sub-levels $M_s = 0$ and ± 1 by the axial field. Thus in a general orientation, the magnetic transitions should not coincide with the F-centre spectrum, and orientation-dependence studies should easily confirm the model of the centre. The negative spin resonance result indicates either that the F_{3c}-centre is not relevant or that the $^3A_2$ ground state is not the one involved. The ^1E ground level is precluded by the stress results, and only $^1A_1$ need be considered as a likely possibility. In magnesium oxide where the electron orbitals are strongly bound to the individual vacancy, the electron–electron interaction is probably small compared with the kinetic and potential energy terms. In this case $^1A_1$ may be preferred, since then the spin eigen functions are orthogonal and the orbital wave functions may freely adjust themselves so as to reduce the kinetic and potential energies. Finally, we note that the parameters A_1 and A_2 describing the magnitude of the orientational shift of the lines under stress are extremely similar in the case of the 6420 Å and 6490 Å lines. As discussed below, the 6490 Å line is due to a vacancy aggregate which does not involve a cation vacancy along the trigonal axis. The similarity in the splitting parameters seems to imply the general similarity in the structure of the defects. Thus it is suggested that the 6420 Å line is due to six electrons localized in three anion vacancies in nearest neighbour sites in the (111) plane (i.e. the F_3 centre). The transitions involved are then $^1A_1 \to {}^1E$.

The 6490 Å has an A ground state and an excited E state (King and Henderson 1966, Ludlow 1968). The only possibilities for this line are those involving the transitions $^3A_2 \to {}^3E$ in the F_{3c} centre and $^2A_1 \to {}^2E$ in the 7-electron F_3 centre. However, spin–orbit interaction in the F_{3c} centre excited state would split the $^3A_2 \leftrightarrow {}^3E$ line into a triplet (not a doublet), eliminating this as a possibility. Consequently we identify this line with a defect containing an odd number of electrons, a ^2A ground state, and an orbitally degenerate ^2E excited state which is split by spin–orbit interaction. The F_3^- centre would appear to have the appropriate properties.

The $^2A_1$ state is of course paramagnetic and in the absence of relaxation time problems inherent in the ^2E state of an R centre, magnetic resonance spectra should be easily observed. Preliminary observations of a paramagnetic trigonal centre have been given by Henderson (1968). This spectrum was observed in crystals which had been irradiated at 600°C and consequently the aggregate centre concentration was of the same order as the F^+ centre concentration. Typical measurements at Q-band for magnetic field orientations in the (100) and (110) planes show that the spectrum may be described by a spin Hamiltonian of the form:

$$\mathcal{H} = \beta[g_\perp(B_x S_x + B_y S_y) + g_\parallel B_z S_z] + A \cdot \vec{I} \cdot \vec{S},$$

in which $g_\perp = 2\cdot 0039$ and $g_\parallel = 2\cdot 0026$. The hyperfine interaction is of some interest since the spectrum indicates that there are two different interactions of importance. Some F_3^- centres will have no magnetic nuclei in the four nearest-neighbour cation sites. Others will have one magnetic ^{25}Mg nucleus (I_α) along the defect axis and equidistant from each of the three anion vacancies. The remainder of the centres will have one Mg^{25} nucleus in one of the cation sites which are equidistant from only two of the anion vacancies (I_β). These two interactions are shown in fig. 39. The total intensity associated with each interaction is related only to the probability of having an Mg^{25} nucleus in the site concerned. Transitions associated with interaction I_α should have only 10% intensity while the total intensity of I_β is of order 25%. This is substantially the case in the spectrum observed. To within experimental error the value of A is about 4 G and is isotropic. Nuclear hyperfine interaction of type I_β is smaller, being only $A_\beta = 3\cdot 2$ G, and is slightly anisotropic ($B_\beta \simeq 0\cdot 8$ G). These results are to be expected for the F_3^- centre and do help to confirm the model predicted on the basis of optical studies. It is emphasized that no one-to-one correlation between the intensity of the optical zero-phonon line and the paramagnetic resonance spectrum has been observed. Nevertheless, the evidence presented here implies that the aggregate character of this and the other centres is assured.

A discordant note should be sounded on the structures suggested on the basis of the stress experiments. In additively coloured crystals, subsequently irradiated with 2 MeV electrons and annealed above 900°C, the orthorhombic and trigonal defects are not present. Since F and F^+ centres diffuse above 900°C it would appear that the structures responsible for these lines must involve cation vacancies and/or interstitials as well as anion vacancies. However the growth rate of the 10445 Å line during electron irradiation is consistent with anion divacancy production (Chen et al. 1970).

6.4.3. Emission studies of trigonal centres

Specimens containing a high concentration of trigonal centres give out a red luminescence when illuminated by a strong source of visible or ultra-violet light (King 1968). At 77°K the luminescence is characterized by an emission band peak at 7200 Å and also sharp lines at 6420 Å and 6490 Å. Weak phonon-assisted structure on the low energy side of the 6490 Å line was also observed. The luminescence is unfortunately rather weak, especially from the 6420 Å line. Further studies of this emission are required in order to confirm the data observed in absorption and also investigate fully such effects as Jahn-Teller distortions expected to occur in the orbitally degenerate E states.

6.5. Other studies in magnesium oxide and calcium oxide

Three other zero-phonon lines have been investigated in annealed crystals of neutron irradiated magnesium oxide. These lines occur at wave lengths of 4232 Å, 5680 Å and 7058 Å. The line at 4232 Å is always too wide (100 cm^{-1}) to allow meaningful uniaxial stress measurements to be made. It may, however, be significant this line is considerably enhanced in crystals which have been doped with chromium prior to neutron irradiation (Ludlow 1966). The 5680 Å line, although relatively narrow, shows no splittings even under stresses approaching 70 Kg/mm^2. The associated defect is extremely stable and the line is still very strong even in crystals annealed to 800°C (King 1967). Splittings under uniaxial stress have been observed for the 7058 Å line. Analysis of the results show that the defect has monoclinic I symmetry with a two-fold axis along the [110] direction and an electric dipole moment along the [112] direction. The result was interpreted as indicating a defect model similar to the N_1 centre in the alkali halides. The most likely possibility involves two anion vacancies along a [112] direction (King and Henderson 1966).

By comparison with magnesium oxide, little work has been reported on the zero-phonon lines in calcium oxide. Neeley (1964) first reported the fine structure on colour centre bands in neutron irradiated calcium oxide but did not comment on the significance of his observations. Henderson and King (1968) have recently analysed the phonon-assisted spectrum associated with the two lines at 5000 Å and 5720 Å. Uniaxial stress results showed also that the line at 5720 Å splits under stress as expected for a centre with orthorhombic symmetry.

We conclude by commenting that no reports of sharp line structure have been published concerning defect states in barium oxide and strontium oxide. It is anticipated that should such observations be made, the preceding work, especially on magnesium oxide, will be invaluable in helping to understand the structure of the defects.

§ 7. The nature of irradiation damage in magnesium oxide

In the previous sections evidence obtained from ESR and optical absorption techniques was utilized to give a fairly detailed description of the nature and magnitude of the neutron-induced damage on the anion sub-lattice. Earlier investigations of the cation displacements using the V^- centre as a probe failed because V^- centres were not produced in neutron-irradiated crystals by subsequent X-irradiation except at very low doses. Although more complex V centres are observed after extended periods of neutron irradiation (Wertz et al. 1961) they are present in such small concentrations that they do not seem to present an accurate picture of the damage on the cation sub-lattice. Several explanations of this behaviour are possible. Firstly, the extremely small Mg^{2+} ion may migrate easily through the crystal to sites from which it was originally displaced. This mechanism for reducing the cation vacancy concentration is not necessarily ruled out by the requirement that some interstitial Mg^{2+} ions must be present in order that interstitial dislocation loops be formed. It is unlikely to operate in the other oxides where larger cations are involved. Another more likely possibility is that damage may be inhomogeneously distributed in the crystal. In regions where the vacancy concentration is high, the presence of F^+ centres will bias the crystal against further positively charged defect species of which the V^- centre is but one. Thus the nature of cation vacancy defects generated by fast neutron irradiation had not been recognized.

This matter has not been fully resolved, the uncertainty being partially associated with the impurities present in most commercially available crystals. In some very early studies of optical absorption Clarke (1957) decomposed the absorption in the V^- band region induced in MgO by fast neutrons into bands at 2·35 eV (5280 Å) and 2·15 eV (5740 Å). With hindsight then it appears that V^- centres were present in those crystals. Most subsequent work has ignored this possibility until very recently. However this is a controversial area which still requires careful work. The most extensive study of V^- centre formation was by Chen and his associates (Chen et al. 1975): they used γ-rays, 2 MeV electrons and reactor neutrons to produce V^- centres. They concluded on the basis of an impressive array of experimental evidences that V^- centres are produced by a photolytic process in which V_{OH}^- centres capture a hole and subsequently decompose to form isolated V^- centres and protons. There is as yet no evidence that V^- centres are formed by the knock-on displacement of Mg^{2+} ions, whether the radiation used be fast electrons, neutrons (Chen et al. 1975), 1·0–4·8 MeV Ne^+ ions (Evans et al. 1972) or 2 MeV Li^+, Na^+ or Tm^{2+} ions (Garrison and Henderson, to be published). Nor does irradiation with energetic metallic ions result in V^- centre formation in CaO. The fate of the protons released by the photolytic reaction proposed by Chen et al. (1975) has not yet been established. V^- centres are observed in n-irradiated CaO although no evidence regarding the involvement of OH$^-$ ions has been reported (Hall 1976).

It is of some importance to establish experimentally the displacement energy for ions in the oxide lattices. Obviously this has not yet been possible for the cation displacement process, although in a report on electron microscope studies of electron-induced damage Sharp and Rumsby (1972) estimate that the Mg^{2+} displacement energy is 64 ± 2 eV. There have been several studies of the O^{2-} ion displacement. Chen and Sibley (1967) first established that the damage mechanism is due to elastic collisions rather than photolysis. Subsequently Chen et al. (1970) investigated the formation of anion vacancies in MgO using fast electrons with energies in the range 0·23–29 MeV. They found that the threshold energy for O^{2-} ion displacement occurs at 0·33 MeV, corresponding to a displacement energy of 60 eV. Hughes (1973) has studied the production of anion vacancy centres in both MgO and CaO using 0·4 MeV and 3 MeV protons at sample temperatures of 15°K, 77°K and 300°K. He finds that the radiation induced vacancy concentration is more than an order of magnitude lower than expected on the basis of simple knock-on collision theory (Kinchin and Pease, 1955). This is in keeping with the results of neutron irradiation studies (Henderson and King 1966). This seemingly implies that any vacancy/interstitial recombination must be correlated within a particular displacement cascade, since random recombination does not lead to a reduced initial growth rate. The ion must therefore be displaced sufficiently far from its vacancy that recombination cannot take place, more interstitials escaping from the critical volume around the vacancy at low temperature than at high temperatures.

Although V^- centres seem to require the presence of OH$^-$ in MgO, this does not necessarily mean that there are no cation vacancies formed during reactor irradiation. Indeed the formation of P^- centres, for which there is EPR evidence, demands the presence of Mg^{2+} vacancies. There is also a wealth of experimental results which have relied heavily upon the assumption that both cations and anions are displaced by fast neutrons. Some of this evidence we will now discuss in detail.

We are also concerned to identify the former occupants of the vacant anion or cation sites. Little direct evidence of point interstitialcies exists. This would occasion no surprise if the damage was in the form of Schottky defects, since Schottky disorder corresponds to the removal of ions from their normal lattice sites to crystal surfaces. The presence of these defects is detected indirectly by measurement of the crystal density, which changes by an amount $\Delta \rho$, related to the concentration of pairs N_P by:

$$-\Delta \rho / \rho = N_P (V_P + \Delta V_P). \tag{54}$$

Here V_P is the volume occupied by an anion/cation pair and ΔV_P represents the relaxation around the vacancy pair. No change occurs in the unit cell size, but only in the macroscopic dimensions of the crystal. When Frenkel disorder persists density changes and unit cell size changes consequent upon the displaced ions being squeezed into the lattice interstices are observed. In this case, and assuming a random distribution of point centres of isotropic dilation, the change in macroscopic volume is equal to the change in unit cell volume. Hence we may write:

$$-\frac{1}{3} \frac{\Delta \rho}{\rho} = \frac{\Delta a}{a} = \tfrac{1}{3}(N_F^- \Delta V^- + N_F^+ \Delta V^+), \tag{55}$$

where $N_F^\pm$ are the numbers of cation/anion Frenkel defects and $\Delta V^\pm$ are the volume changes per Frenkel defect, including the dilatation around the vacancy and interstitial. When both Frenkel and Schottky defects are present, a suitable summation of the above equations is involved. We observe that sufficiently accurate measurement of the growth of the unit cell ($\Delta a/a$) and macroscopic dimensions of the crystal ($\Delta \rho/3\rho$) will indicate unequivocally the type of defect present. It does not identify the particular form of point defect.

7.1. *Evidence of point interstitials*

In figs. 40 and 41 the defect concentrations are compared with crystal dimensional changes for neutron doses up to 10^{20} nvt. Evidently $\Delta a/a$ and $\Delta \rho/3\rho$ are almost identical at doses up to 3×10^{19} nvt whereas above 5×10^{19} nvt $\Delta \rho/3\rho$ is measurably greater than $\Delta a/a$. At doses greater than 3×10^{19} nvt the F^+ centre concentration reaches a saturation level of about 10^{19} F^+ cm^{-3}; aggregate defects are also increasingly apparent at this dose. The excellent accord between $\Delta a/a$ and $\Delta \rho/3\rho$ at low doses indicates the predominance of Frenkel disorder in these crystals, according to the foregoing arguments. This does not preclude interstitial clustering on a small scale: it does imply that if small clusters are formed, the dilation per interstitial ion remains unaffected by its presence in the cluster.

Since we cannot separate the contributions to the dimensional changes by anion and cation Frenkel defects we simplify eqn. (55) by writing

$$\frac{\Delta a}{a} = 2\alpha \left(\frac{N_F}{N_0} \right)$$

where α is the mean linear dilatation per Frenkel pair, N_0 is the total number of ions per cm^3 and N_F is the F^+ centre concentration (and consequently the anion Frenkel defect concentration). The factor 2 arises because we assume that neutron irradiation produces equal numbers of Frenkel defects on anion and cation sub-lattices. Differentiating with respect to dose gives

$$\frac{d(\Delta a/a)}{dD} = \frac{2\alpha}{N_0} \frac{dN_F}{dD}. \qquad (56)$$

Thus by comparison of the slopes of figs. 40 and 41 a value of α may be obtained. Henderson and Bowen (1971) find that in the range 10^{17}–10^{19} nvt α is in the range 2·2–2·9. Thus the mean volume dilatation per Frenkel pair is 6·6–8·7 atomic volumes. (Note one Frenkel pair comprises Mg^{2+} and O^{2-} interstitials and vacancies.)

Two important differences are apparent between the results of Henderson and Bowen (1971) and the earlier ones of Hickman and Walker (1965). Firstly, Hickman and Walker find that at doses up to 7×10^{19} nvt $\Delta a/a$ exceeds $\Delta \rho/3\rho$ by a small but significant amount; in addition, their observed growth parameters are always less than those observed by Henderson and Bowen. It is concluded that $\Delta \rho/3\rho < \Delta a/a$ is associated with preferential annealing of vacancies resulting in a small excess of interstitials ($\sim 10\%$) being present (Hickman and Walker 1965). This and the lower values of both $\Delta a/a$ and $\Delta \rho/3\rho$ may be associated with different ambient temperatures of the reactors. Hickman and Walker (1965) used irradiation temperatures near 100°C whilst Henderson and Bowen used specimens which had attained temperatures of only 40 ± 10°C. Presumably, higher temperatures facilitate recombination processes and reduce the defect concentration by a small amount. Generally, the observation that the macroscopic crystal volume and unit cell volume are equally affected by small doses of fast neutrons, strongly suggests the presence of equal numbers of essentially isolated vacancies and interstitials.

Fig. 40

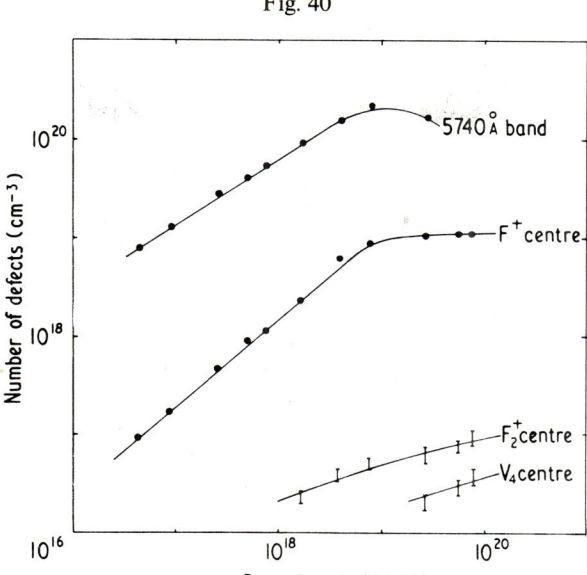

The concentration of various defects in MgO single crystals after reactor irradiation to fast neutron doses in the range 10^{16}–10^{21} nvt. (After Henderson and Bowen 1971.)

Fig. 41

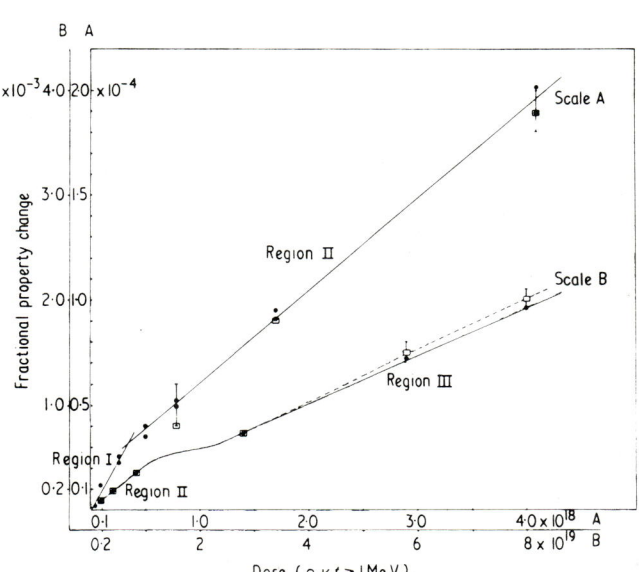

The fractional change in lattice parameter $\Delta a/a$ and crystal density $\Delta\rho/3\rho$ plotted as a function of dose up to 4×10^{18} nvt (scale A) and 8×10^{19} nvt (scale B). The scales for both ordinate and abscissa have been changed by the same factor $\times 20$ in converting scale A to scale B. Full curve with full squares, $\Delta a/a$; broken curve with open squares, $\frac{1}{3}\Delta\rho/\rho$. (After Henderson and Bowen 1971.)

At higher doses ($>5 \times 10^{19}$ nvt) $\Delta\rho/3\rho$ is greater than $\Delta a/a$ by an amount which increases with increasing fast neutron dose (fig. 42). At such dose levels the saturation concentration of F^+ centres is less than half the concentration of oxygen ions present in the clustered defects observed by Groves and Kelly (1963) and Bowen and Clarke (1964). Henderson and King suggested that this results from small vacancy clusters being formed. That $\Delta\rho/3\rho$ exceeds $\Delta a/a$ is effected by interstitial clusters being formed. According to eqn. (54) these do not contribute to X-ray growth. These Schottky defects may be observed in suitably thinned crystals by electron microscopy (Bowen et al. 1962, Groves and Kelly 1963, Bowen and Clarke 1964). They appear in the electron microscope as black dots $\sim 50°$ in diameter, that can be tilted in and out of contrast by varying the angle between the foil and electron beam. Annealing the samples above 800°C coarsens the dots such they can readily be recognized as dislocation loops (Groves and Kelly 1963). Detailed analysis of the contrast conditions for observation of these loops (Groves and Kelly 1962, Ashby and Brown 1963), showed them to be interstitial loops lying in the (110) lattice plane. Furthermore, no stacking faults appear within the loops so that during formation perfect (110) stacking sequences have been maintained. To satisfy this condition two layers of interstitial atoms would have to be inserted in the (110) sequence, alternate rows consisting of equal numbers of O^{2-} and Mg^{2+} ions. An important consideration here is that O^{2-} ions are involved in such dislocation loops; since this ion is the primary source of electrons in the formation of F-like centres, it is not surprising that the F^+ centres rather than the F-centre appears to be the stable anion defect in neutron-irradiated crystals.

It is conceivable that impurity ions may act as an alternative source of electrons.

Fig. 42

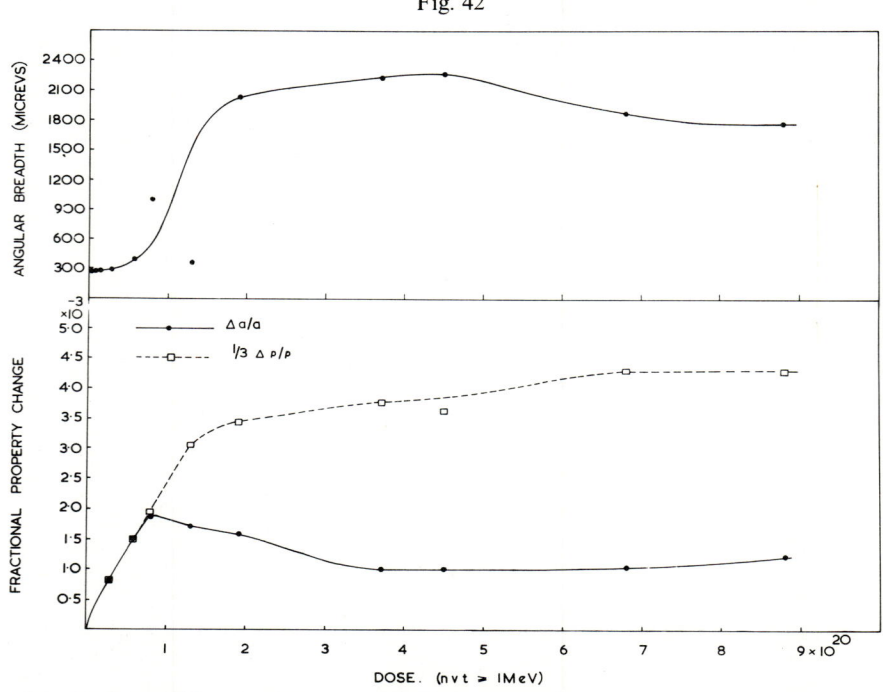

Showing the large difference between $\Delta a/a$ (full curve) and $\Delta\rho/3\rho$ (broken curve) and doses $>5 \times 10$ nvt. (After Henderson et al. 1971 c.)

Observations on crystals varying in Fe concentration by 3 orders of magnitude show that variable valency ions play only a minor role in providing electrons for trapping at oxygen ions vacancies. One impurity which does have a profound influence on defect production is hydrogen. The association of OH^- with vacancies and the presence of $Mg(OH)_2$ has already been discussed at some length in Chapter 4. When MgO crystals are annealed in hydrogen at 1200°C prior to neutron irradiation one observes large differences in dimensional changes and colour centre concentrations relative to unannealed crystals (Henderson et al. 1971 c). The effect of annealing in H_2 is to produce large cavities in the bulk of the crystal with sides parallel to $\langle 110 \rangle$ directions and lenticular in cross section. Implicit in the cavitation mechanism is the fact that the H_2 which fills the cavities is internal to the crystal. Thus cavities are produced irrespective of whether H_2 or CO is the reducing medium (Briggs and Bowen 1968). The lower defect concentrations induced by reactor irradiation are due to vacancy centres diffusing to these voids and become lost as radiation products. Henderson et al. (1971 c) deduce that a vacancy formed within a radius of 7×10^{-5} m around the vacancy will be absorbed by the vacancy, unsurprising in view of the extensive strain field observed in the neighbourhood of the cavity. This mechanism does not involve the interstitials, thus measurements of dimensional changes in annealed and as-received crystals are very similar at least up to a dose of $\sim 2 \times 10^{18}$ nvt. In the dose range 2×10^{18} nvt to 6×10^{19} nvt, however, $\Delta a/a$ is slightly in excess of $\Delta \rho / 3\rho$. This is attributed to an excess of interstitials over vacancies, as a consequence of vacancies becoming trapped at the voids. Such an explanation may also obtain in the case of the results of Hickman and Walker (1965); frequently as-grown crystals of MgO contain voids which are visible to the naked eye. The contraction per vacancy is estimated to be about one atomic volume (Henderson et al. 1971).

It may appear surprising that we assume equal numbers of Frenkel defects on the anion and cation sublattices. There are good reasons for so doing. First we know that on annealing reactor irradiated crystals at temperatures in excess of 300°C results in F^+ centres being converted to P^- centres. This process increases up to about 900°C after which at higher temperatures P^- centres are lost by aggregation to higher order defects. The formation of voids is believed to occur by vacancy-pair aggregation as is discussed below.

7.2. Annealing studies and cavity formation

Magnesium oxide annealed at elevated temperatures, subsequent to fast neutron irradiation, recovers both its lattice parameter and macroscopic density. The recovery of $\Delta a/a$ is more rapid than $\Delta \rho / 3\rho$ (Bowen and Clarke 1964, Hickman and Walker 1965). For both $\Delta \rho / 3\rho$ and $\Delta a/a$ recovery is dose-dependent, higher temperatures being required for complete recovery of the macroscopic density in crystals irradiated at higher doses. Surprisingly, the lattice parameter falls below the unirradiated value at some temperatures in samples irradiated to fast neutron doses above 3×10^{19} nvt. At this dose, the negative change in the lattice parameter is observed at ~ 1050°C: complete recovery of the lattice parameter is delayed to a temperature of 1500°C. For samples irradiated at $8 \cdot 8 \times 10^{20}$ nvt complete recovery of the lattice parameter is not achieved until ~ 1800°C.

The recovery of lattice parameter during annealing is consequent upon several processes occurring at the elevated temperature. Vacancy–interstitial annihilation does take place but the observation of P^- centres at 900°C and voids (see below) at much higher temperatures convinces us that this is not the most prominent mechanism. Interstitial migration to existing dislocation loops would seem to be taking place. This is evident from the growth of small clusters represented by the black dots in electron micrographs; these grow into well defined dislocation loops above 800°C. Once present in the loops, interstitial ions contribute only to the macroscopic growth, and consequently both $\Delta a/a$ and $\Delta \rho/3\rho$ decrease, with $\Delta \rho/3\rho$ decreasing less rapidly. Not only do the loops increase in size, but they also decrease in concentration. This occurs because dislocation loops with [110] Burgers vectors are confined to (110) lattice planes and may glide on the cylinder defined by the loop perimeter and Burgers vector, and so amalgamate with other loops lying on the same slip plane. By a similar process, thermally activated glissile dislocations may interact with dislocations lying on other slip planes, to produce tangled networks of dislocations (Bowen and Clarke 1964). This situation obtains at higher doses and the resistance to annealing of high dose samples is related to the stability of these dislocation tangles.

That lattice parameter values below the unirradiated value occur at high temperature argues that vacancy aggregates are present which cause a net lattice contraction. Such vacancy aggregates are evidently too small to observe in the electron microscope, although Bowen and Clarke (1964) did observe small defect clusters in irradiated samples annealed at temperatures of 1500°C (where all loops had annealed out). They did not specifically associate these defects with vacancies. Such evidence was adduced by Morgan and Bowen (1967) who observed cuboidal voids over a range of doses, irradiation temperatures and annealing temperatures. These cavities are bounded by (100) lattice planes, and vary in size. The precise mechanism of formation is not known but the precursor may be the P^- centre. Such centres become increasingly less stable above 900°C, although copious numbers are present at this temperature. The P^- centre is easily ionized, and the resulting vacancy pair becomes mobile above 900°C, whence it migrates in order to coalesce with other pairs. This is consistent with the experimental evidence that P^- centres cannot be regenerated by X-rays in crystals annealed above 900°C. Whatever the nature of the cluster formed by the migratory anion–cation vacancy pairs it is immobile below 1500°C, since the cuboidal cavities are not formed below this temperature. The clusters apparently have an associated contractive dilation which causes the negative change in lattice parameter. Martin (1968) has deduced from cold neutron scattering measurements that clusters containing up to ~ 100 vacancies are present in samples irradiated and annealed under the identical condition used by Morgan and Bowen (1967). The cuboidal voids are not observed in crystals irradiated with less than $\sim 10^{20}$ nvt and a supersaturation of vacancies is evidently required for their formation (Morgan and Bowen 1967). When the clusters do condense to form voids their dilations are removed and the lattice parameter increases. The temperature at which the negative change in parameter begins to recover correlates well with that above which the cavities are observed in the electron microscope (Briggs and Bowen 1968). It has been established that these cavities contain the gaseous transmutation products Ne and He at pressures not exceeding 10 atmospheres. Morgan and Bowen (1967) concluded from studies of the gas released by crushing crystals which had been annealed at 1800°C for varying periods, that the majority of gas atoms are initially trapped at irradiation-induced dislocation lines. Diffusion of appreciable numbers of gas atoms into the cavities occurred only when the annealing period was sufficiently long to remove the majority of dislocations.

It is of interest to note that unirradiated MgO crystals that have undergone extensive heating in H_2 at about 1300°C show lenticular shaped cavities. These cavities contain H_2 under pressures estimated to be of order 250 atm. (Briggs and Bowen 1968). Precisely the same behaviour is observed whether the crystal is heated in D_2 or CO. It should also be recorded that heating MgO crystals for 3 weeks at 1500°C is ineffective in producing any detectable OD bands. Hence the source of H_2 must have been internal. This is consistent with the cavities being observed in samples which show a strong OH^- absorption.

§ 8. Oxide surfaces and catalysis

8.1. *ESR studies of defects* In *and* On *polycrystalline oxides*

Surface studies are very much in vogue, largely because of recent development of the necessary experimental techniques. However there is also a new awareness that surface properties modify the properties of technological devices: to this must be added the former knowledge of such processes as corrosion, fracture, lubrication and crystal growth. In semiconductor devices we are usually concerned with solid–solid surfaces, this Chapter concentrates on effects at solid-vacuum interfaces.

The first point to be made is that the properties of the atomic planes close to a solid-vacuum interface are modified. For ionic surfaces ionic displacements occur, the magnitude of which are determined by ionic size and polarizability. Theoretical and experimental work has been reported for MgO but, insofar as the author is aware, not for other alkaline earth oxides. The surface Mg^{2+} ions are predicted to relax inwards by a few percent so that the mean lattice spacing of finely divided MgO powder is also reduced (Anderson and Scholtz 1968). The experimental evidence using small cubic crystallites (MgO smoke) shows average ionic displacements which are in keeping with the theoretical values. Such changes in the average ionic spacing reduce the surface energy σ, which is agreed to be in the range 1·0–1·2 J m^{-2} for the (100) surface in MgO. In this material, where the distortion is significant well into the crystal, σ is reduced by about 25%. One further important effect is that the surface may appreciably modify the band gap. This has been reported by Nelson and Hale (1971) for MgO, CaO and SrO surfaces, in which the presence of the surface produces a local band gap at lower energy than that in large single crystals. Surface phonon modes, so important in acoustoelectronic device materials such as $LiNbO_3$, have not been observed in the alkaline earth oxides.

The surface properties of MgO, CaO and SrO have been much studied largely using spectroscopic methods such as ESR, infra-red spectroscopy and diffuse reflectance. In addition to the expected range of intrinsic surface states, numerous impurity adsorbed species have been observed. The experimental evidence for these species is reviewed below, and where important discussed in terms of the surface activity and catalysis. There are similarities in the structure of surface and bulk defects and a notation based upon that used for bulk defects (Henderson and Wertz 1968, Sonder and Sibley 1971) has been proposed by Tench et al. (1972). A subscript 's' is used for intrinsic defects present in the first surface layer of a crystal. Thus the notation F_s^+ indicates an electron trapped in an O^{2-} vacancy at the surface and V_s^- refers to a hole trapped on an O^{2-} ion adjacent to a Mg^{2+} ion vacancy at the surface. Note that the presence of the surface reduces the symmetry around the vacancy from cubic to tetragonal or lower symmetry. In a number of cases impurities are associated with a defect e.g. $F_s^+(OH^-)$ indicates an OH^- ion in the neighbourhood of an F_s^+

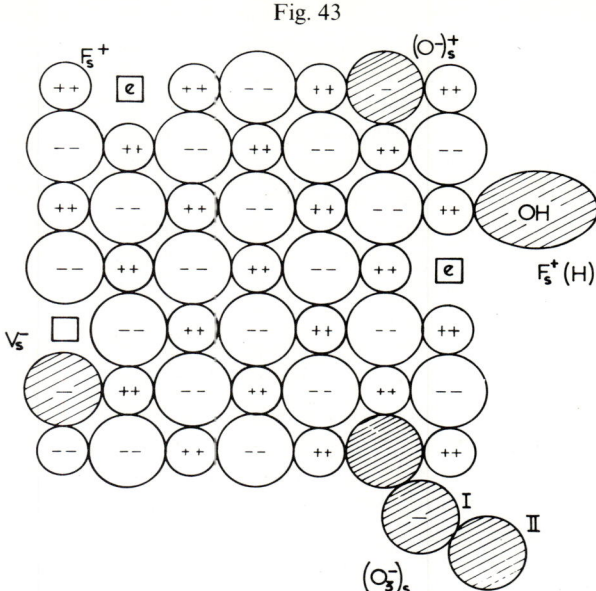

Fig. 43

Surface defects in oxides: the nomenclature is that proposed by Tench et al. (1972).

centre. Several molecular species have been identified: such species are written in brackets and if adsorbed at a specific lattice site then the effective charge is added as a superscript e.g. $(O^-)_s^+$ is an O^- ion occupying an oxygen vacancy on the surface. The structures of some of these surface defects are shown in fig. 43.

One aim of surface studies is to understand the mechanism of heterogeneous catalysis, in particular the influence which the solid state characteristics of the catalyst may have upon the nature and the properties of the adsorbed species. Surfaces containing paramagnetic centres act as catalytic agents in a variety of chemical reactions, and transition metal compounds, which are usually paramagnetic, are used extensively as catalysts. Even in non-magnetic systems chemisorption of molecules on insulating surfaces may result in the formation of paramagnetic radicals which may be amenable to study by magnetic resonance spectroscopy. In such cases ESR identifies the nature of the adsorbed species and gives information about surface bonding and the geometrical configuration of the adatoms on the adsorbing surface. The technique is limited to intrinsically paramagnetic atoms, molecules or defects: it has often been useful to irradiate the finely divided powder sample before or after adsorption of gases to produce centres which may be detected by ESR. Ultra-violet light, X- and γ-rays, energetic electrons and protons and reactor neutrons have been used by numerous authors.

Optical spectroscopy has also been found useful in characterizing the nature of oxide surfaces. For measurements in the near ultra-violet and visible spectrum transmission measurements are of little value because scattering is the dominant feature. However diffuse reflectance spectroscopy has been used in many studies of powdered materials. The reflectance may be used to obtain the absorption coefficient (μ) using the relation

$$\frac{\mu}{s} = \frac{(1-R)^2}{2R} \tag{56}$$

where R is the percent reflectance (Kortüm 1969). However, care must be exercised in using eqn. (56) since the scattering coefficient, s, depends upon particle size and wave length and is not readily computed with accuracy. However, as Nelson and Hale (1971) point out the technique simultaneously monitors absorption of the incident radiation by bulk defects and by surface species, and has proved particularly suitable for study of such oxides as MgO, CaO and SrO. Infra-red spectroscopy is of little value except when molecular species are adsorbed on surfaces. Since in catalysis one is interested in reactions involving a wide range of inorganic and organic molecules it has been applied widely, especially in the range 1000–$10\,000\,\mathrm{cm}^{-1}$. Conventional transmission spectroscopy has been the order of the day, the absorption of radiation by compressed discs of oxide being measured. However, the importance of infra-red reflectance spectroscopy is now being recognized (Kunath and Reklat 1971).

The ESR technique has given important information on the following types of defects in oxide powders:

(a) Transition-metal or rare-earth ions as substitutional impurities. Most of the ions detected show only a small anisotropy in the ESR spectrum. Ions which have been observed in MgO powders include V^{2+}, Cr^{3+}, Mn^{2+}, Mn^{4+}, Fe^{3+}, Co^{2+} and Cu^{2+}. For CaO powders, the V^{2+} and Co^{2+} ions are not seen, but Gd^{3+} is. For SrO powders, Mn^{2+}, Fe^{3+} and Cr^{3+} and Gd^{3+} are seen (Auzins et al. 1963). These defects will not be considered further.

(b) Electrons trapped at anion vacancies *within* crystallites, i.e. F^+ and P^- centres.

(c) Trapped-hole centres within crystallites.

(d) Surface centres consisting of electrons trapped at anion vacancies or vacancy pairs. There is some evidence for the occurrence of surface analogues of V^- centres.

(e) ESR studies of paramagnetic species adsorbed on the surface of MgO. These centres include CO_2^-, NO, O_2^- and CO.

Normally, single crystals are much to be preferred as compared with powders. However, for some studies, the use of powders or hot-pressed aggregates gives significant advantages of convenience. These include:

(1) Ability to dope with precious agents, such as ^{43}Ca; conventional methods of single-crystal preparation would be prohibitively wasteful.

(2) Ability to dope with aqueous solutions and to achieve higher levels of doping than are readily realized in crystals. In at least one instance, this has aided the identification of a spectrum (V_F centre, § 5) seen in a single crystal.

(3) Rapid generation of F^+ centres (and of P^- centres by subsequent heating) in any of the oxides without generation of high levels of radioactivity.

(4) Ability to detect surface centres, to observe their disappearance during absorption of gases and to detect new paramagnetic adsorbed species; concentrations of these defects are far too low to permit detection on single crystals with existing ESR techniques.

8.1.1. *Formation of F^+ centres in oxide powders*

Comparative studies of g factors of F^+ centres in all of the oxides, sulphides and selenides of the alkaline earth metals have been greatly facilitated by the discovery that F^+ centres are generated by grinding these materials (Wertz et al. 1961, Walters and Estle 1961, Auzins et al. 1963). Heating at $700°K$ followed by X-irradiation produces P^- centres, as in single crystals of MgO, CaO or SrO. Grinding induces dislocations and causes them to move considerable distances within a crystallite. The presence of F^+ centres in ground powders suggests that *single* anion vacancies are dissociated from the dislocations as they move. Population of these vacancies with an electron requires thermal dissociation of electrons from an impurity ion or from the lattice oxide ions during the grinding process. The F^+ centre signal intensities are often enhanced by X-irradiation after grinding. The production of P^- centres by heating at 70 °K and subsequent X-irradiation is suggestive of cation vacancy migration.

Neutron irradiation of powders of the alkaline earth oxides produces F^+ centres as in single crystals. Enrichment of CaO with ^{43}Ca allows one to observe hyperfine splitting from the F^+ centre: ^{43}Ca hyperfine parameters determined using such powders are closely similar to those obtained from single crystal measurements (table 10) as are data for the other oxide powders. The axial parameter b of the hyperfine interaction is obtained from the separation of turning points corresponding to a given value of M_I, viz. $|3b|M_I$. For SrO, only the lines corresponding to $M_I = \pm\frac{1}{2}$, $\pm\frac{3}{2}$ were observed in powders. Separate values of the hyperfine parameters for ^{135}Ba and ^{137}Ba in polycrystalline BaO were determined with the aid of a signal-averaging system (Tench and Nelson 1967 a). Only one overlapping quartet had been reported in the earlier single-crystal study of BaO (Carson *et al.* 1959).

If MgO or CaO powders are neutron irradiated *in vacuo* and oxygen is then admitted, the concentration of F^+ centres is found to diminish slowly (Nelson and Tench 1964, Nelson *et al.* 1967 a). This is connected with chemisorption of the oxygen (§ 8.1.4); surface analogues of the F^+ centre (§ 8.1.3) disappear far more rapidly.

8.1.2. *Trapped hole centres in magnesium oxide powders*

The narrow line widths of the V^-, V_{OH} and the V_F centres in single crystals of MgO (<0·5 G) makes it possible to observe both the parallel and perpendicular line components in powders if the concentration of Fe^{3+} is sufficiently low (Auzins *et al.* 1963, Wertz and Auzins 1965). The V^- centre shows precisely the line shape expected for a system with axial symmetry (Ibers and Swalen 1962, Weil and Hecht 1963, Kneubühl 1960). X-irradiated MgO powders made from the hydroxide show a splitting of the intense $g_\perp$ line of the V_{OH} centre. Powders treated with HF solution, heated at 1500°K and then X-irradiated show a smaller splitting of the perpendicular line. This behaviour was helpful in implicating a substitutional fluorine atom as being responsible for hyperfine splitting of the lines of a V_1-like spectrum (§ 4).

8.1.3. *Trapped electron centres at surfaces*

Neutron irradiation in vacuo of magnesium oxide powders produce a distinctive red–blue colour which disappears very rapidly on exposure to oxygen. Before admission of oxygen, the ESR spectrum shows an asymmetric line with $g_{av} = 2·0007$ (Nelson and Tench 1964, Nelson *et al.* 1967). This F_s^+ centre line disappears along with the colour, and a fast chemisorption of oxygen occurs. For calcium oxide powders neutron irradiated *in vacuo*, a blue–green colour is observed; its presence is correlated with an ESR spectrum characteristic of axial symmetry, with $g_\parallel = 1·9992$ and $g_\perp = 1·9969$. It is likewise designated as an F_s^+ centre. The colour and the ESR lines disappear rapidly upon admission of oxygen. In an attempt to detect hyperfine interaction with electrons trapped in the surface region, magnesium oxide powders were prepared from 80% enriched ^{25}Mg. The observed line has a width of 15 G, without any indication of resolved structure. A calcium oxide sample enriched to 2% in ^{43}Ca shows a hyperfine octet from the F^+ centre, but no detectable hyperfine splitting from the F_s^+ centre. A sample enriched to 44% ^{43}Ca showed no resolved F_s^+ spectrum but two sets of hyperfine lines superimposed on a 20 G wide line. The latter disappears immediately on admission of oxygen; it is presumably an inhomogeneously broadened line due to unresolved hyperfine structure of the F_s^+ centre. A weak F_s^+-like signal with $g_{av} = 1·9798$ is seen from SrO.

For either MgO or CaO it is proposed that the F_s^+ centre represents an electron trapped in a surface anion vacancy (Nelson et al. 1967). The negative Δg values are presumed to result from the effect of increased polarization of five cations as compared with the six of the F^+ centre. Adsorption of O_2 is presumed to be the result of addition of an F_s^+-centre electron to O_2 to give O_2^-, which then may occupy the anion vacancy.

A different surface centre with g components, and microwave saturation behaviour also appropriate to F^+-type centres is formed upon ultra-violet or γ-irradiation of MgO. This has been termed the S' centre (Lunsford and Jayne 1965). In contrast with the axial symmetry of the F_s^+ centre, the S' centre appears to have rhombic symmetry with $g_x = 2·0015$, $g_y = 2·0015$ and $g_z = 2·0005$ (Lunsford and Jayne 1965, Nelson et al. 1967). It is presumed that the S' centre is a surface analogue of the P^- centre in the alkaline-earth oxides (Wertz et al. 1961). The g components measured at 35 GHz for the P^- centre in MgO are $g_\perp = 2·004$ and $g_\parallel = 2·0012$ and in CaO, $g_\parallel = 1·9995$ and $g_\perp = 1·9980$ (To et al. (1971)). Similarity of the g factors of the P^- and the F_s^+ centres and of their behaviour towards oxygen suggests that each has an electron in an anion vacancy in the surface. However, the P_s^- centre is presumed to have a cation vacancy associated with the anion vacancy in the surface.

8.1.4. Adsorbed paramagnetic species on magnesium oxide powders

MgO powders which have been outgassed at 1075°K and irradiated with 5 eV light show a rapid adsorption of oxygen and give a new ESR spectrum. The principal g components of the spectrum are $g_{xx} = 2·077$, $g_{yy} = 2·0011$ and $g_{zz} = 2·0073$. Comparison of these values with those from O_2^- in KCl, in $CaCO_3$ or in NaO_2 (table 21) suggests that the adsorbed species responsible for the ESR signal is O_2^- (Lunsford and Jayne 1966). A similar signal is obtained by admission of oxygen to which has first been neutron irradiated in vacuo and then heated at 575°K (Tench and Nelson 1966). While the adsorbed species is a surface defect, its g components correspond better with those of O_2^- in NaO_2 than in KCl. It is postulated that the 5 eV irradiation leads to the formation of F_s^+ centres, while neutron irradiation produces F^+ centres also. The former are eliminated upon admission of O_2, while the number of the latter are diminished.

Similarly, CO_2 is chemisorbed on MgO powders pretreated to give F_s^+ and F^+ centres; the g components are: $g_{xx} = 2·0020$, $g_{yy} = 1·9974$ and $g_{zz} = 2·0017$. These correspond very well with those of CO_2^- in $CaCO_3$ or in $HCOONa$ (sodium formate), and hence this identification seems to be confirmed (Lunsford and Jayne 1965). The CO_2^- on the surface must thus be rigid—on the ESR time scale.

Unirradiated MgO powders which have been outgassed at 1175°K and then exposed to nitrobenzene ($C_6H_5NO_2$ = NB) vapours give an ESR spectrum attributed to the NB ion (Tench and Nelson 1967 a, b). In this case, an O^{2-} ion with only a

Table 21. ESR parameters for species adsorbed on MgO powders* and reference species in single crystals

Host	Adsorbate	Paramagnetic species	g_{xx}	g_{yy}	g_{zz}	References
MgO*	O_2	O_2^-	2·077	2·0011	2·0073	Lunsford and Jayne (1966 a)
						Tench and Nelson (1966)†
NaO_2		O_2^-	2·175	2·000	2·000	Bennett et al. (1956)
KCl		O_2^-	2·4359	1·9512	1·9551	Känsig and Cohen (1959)
MgO*	CO_2	CO_2^-	2·0020	1·9974	2·0017	Lunsford and Jayne (1965)
$CaCO_3$		CO_2^-	2·0032	1·9973	2·0016	Ovenall and Whiffen (1961)
HCOONa		CO_2^-	2·0032	1·9975	2·0014	Ovenall and Whiffen (1961)
MgO*	CO	'CO'	2·0055		2·0021	Lunsford and Jayne (1966 b)
MgO*	$C_6H_5NO_2$	$C_6H_5NO_2$	2·006			Tench and Nelson (1967)

† Using gases enriched with 58% ^{17}O, Tench and Holroyd (1968) observe that $A_{xx} = 77 \pm 2$ G, $A_{yy} = 0 \pm 4$ G and $A_{zz} = 15 \pm 2$ G.^{e}

small coordination number is supposed to donate an electron to an acceptor of sufficiently high electron affinity. For NB, the latter is taken as 0·7 eV, while that of O_1^- is 0·46 eV (Branscomb 1962). The reaction

$$O^{2-} \rightarrow O^- + e \quad \text{with} \quad \Delta U = -6·5 \text{ eV}$$

is presumed to take place at oxygen sites for which the stabilization energy resulting from coulombic interaction does not exceed the electron affinity of the adsorbed species. Thus it is presumed that favourably surrounded O^{2-} ions on the surface of MgO may act as electron donors to molecules with electron affinities of 0·7 eV but not to those for which it is less than 0·5 eV.

When NO is adsorbed on MgO powders of area 150 to 250 $m^2 g^{-1}$, distinctly different ESR spectra are seen at 93° and at 296°K (Lunsford 1967). That at 93°K is attributed to adsorbed NO which has its orbital angular momentum largely quenched, since $g_{xx} = g_{yy} = 1·995$ and $g_{zz} = 1·90$. The ^{14}NO species shows well-defined hyperfine splitting of the perpendicular component into a triplet, with $a_{xx} = 33$ G; the upper limit of a_{zz} is given as < 10 G. The comparable ^{15}NO species correspondingly shows a doublet splitting with $a_{xx} = 47$ G and $a_{zz} < 10$ G. For the ion N_2^- in KN_3, the principal g values are: $g_{xx} = 2·0027$, $g_{yy} = 2·008$ and $g_{zz} = 1·9832$. The principal values of the splittings for N_2^- in KN_3 are $a_{xx} = 12(4)$, $a_{yy} = 3·8(2)$ and $a_{zz} = 4(2)$ (Horst et al. 1962). It has been suggested that the g components are sufficiently similar to those of N_2^- to indicate an isoelectronic species, viz. NO. If the π-antibonding levels are split by a large amount, the parallel component of the g tensor should be given by (Mergerian and Marshall 1962):

$$g_{zz} = g_e - 2(\tfrac{1}{2}\lambda)^2 / [\Delta^2 + (\tfrac{1}{2}\lambda)^2]^{1/2}.$$

Here λ is the spin–orbit coupling constant, which represents for the free molecule the separation of the $^2\pi_{1/2}$ and $^2\pi_{3/2}$ states (121 cm^{-1}); Δ is the splitting by the crystal field at the surface of the MgO of the π^* levels. From the experimental value of g_{zz}, $\lambda/\Delta = 0·11$.

The ESR spectra of ^{14}NO and ^{15}NO adsorbed on MgO at 296°K show a very different behaviour. Both powder spectra are dominated by a large hyperfine splitting, viz. $a_{zz} = 43$ G for ^{14}NO and 61 G for ^{15}NO. For both species, $a_{xx} \approx a_{yy} \approx 0$. For ^{14}NO, $g_{xx} = 2 \cdot 0068 = g_{yy}$, $g_{zz} = 2 \cdot 0025$. The similarity with the parameters of NO_2^{2-} (Mergerian and Marshall 1962, Symons 1962, Morton 1964) is taken as evidence that NO at room temperature reacts with an oxide ion (Lunsford 1967).

The F_s^+ centre differs from the F^+ centre mainly in the axial symmetry of the g tensor and the magnitude of the A tensor. As in the case of bulk P^- centres (To et al. 1968) the electronic charge is centred closer to the axial ^{25}Mg so that the hyperfine constant for this ion is enhanced. A simple variant of the F_s^+ centre is observed when MgO powder prepared by thermal decomposition of $Mg(OH)_2$ or $MgCO_3$ respectively is γ or ultra-violet-irradiated in vacuo or in the presence of an overpressure of 1–6 kN m^{-2} of hydrogen. A spectrum is then observed due to the $F_s^+(H)$ centre, a surface defect involving F_s^+ with a nearby hydrogen: the ^{25}Mg hyperfine structure of this $F_s^+(H)$ centre shows an unusual temperature dependence (Tench and Nelson 1968, Smith and Tench 1968). These centres are formed more readily when the irradiation is performed at low temperature. Nitrous oxide has been shown to react

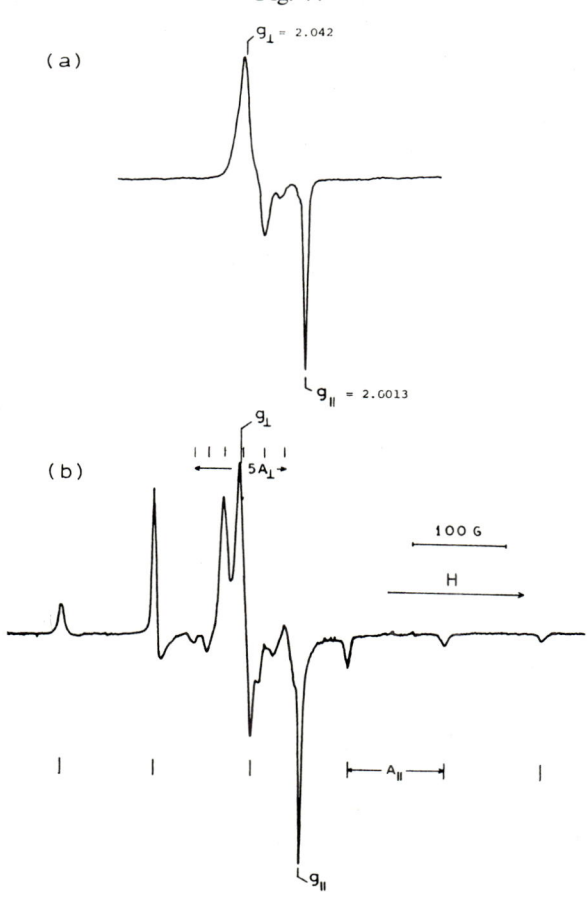

Fig. 44

The spectrum (a) of $(^{16}O^-)_s^+$ centres in MgO and the effect of enrichment with 71·9% ^{17}O (spectrum (b)). (After Wong and Lunsford 1971.)

with $F_s^+(H)$ centres to give $(O^-)_s^+$ ions at the anion vacancy (Wong and Lunsford 1971, Tench et al. 1972). The O^- ion has often been suggested to be an intermediate species in surface reactions, so that its identification is important. The spectrum observed when N_2O reacts with $F_s^+(H)$ centres depends upon the reaction temperature. At $\sim 77°K$ the spectrum is essentially axial (Williamson et al. 1971, Wong and Lunsford 1971) as can be seen in fig. 44: the evidence of the spectrum obtained using $N_2^{17}O$ enriched with 71·9% ^{17}O convincingly demonstrates that a monoatomic species is involved. The $(O^-)_s^+$ centre ESR spectrum is tetragonally symmetric with $g_\perp = 2·042$, $g_\parallel = 2·0013$, $A_\perp = 1·95 \pm 0·01$ mT, $A_\parallel = 10·32 \pm 0·01$ mT and $Q_\perp = 0·13 \pm 0·02$ mT. These results confirm that the hole remains essentially in the $2p_z$ orbital, with only a small degree of covalent admixture of $2s$ character. Furthermore the accurate tetragonal symmetry apparently confirms that the O^- ion is trapped in an O^{2-} vacancy on the surface (fig. 43). The tetragonal axis is, of course, the [100] direction passing through the O^- ion perpendicular to the plane of the surface.

Reaction of N_2O with $F_s^+(H)$ centres at 300°K reveals three different trapping sites for the O^- ion, one of which is that referred to above (Tench et al. 1972). The two other sites have orthorhombic symmetry, the deviation from tetragonality apparently being due to some surface impurity such as the hydroxyl ion.

MgO surfaces containing $(O^-)_s^+$ ions (formed using the $N_2O/F_s^+(H)$ method) react with O_2 to produce the molecular species $(O_3^-)_s$. The observed ESR spectrum is too complex to originate from a single species with only g-tensor anisotropy; a small hyperfine splitting due to a nearby proton is observed for the major species with a g-tensor of $g_1 = 2·0148$, $g_2 = 2·0121$ and $g_3 = 2·0020$. With enriched $^{17}O_2$ a characteristic hyperfine structure is observed, which for a single nucleus of ^{17}O gives two hyperfine sextets characterized by the tensor components 302, 42 and 42 (± 5 MHz) and approximately 196, 28 and 28 (MHz). The larger of these tensors is contributed by $(O_3^-)_s$ centres in which the central atom is the isotope ^{17}O (see fig. 43). Assuming no isotopic exchange between ($^{16}O_s^-$) and $^{17}O_2$ then the smaller tensor components are assumed due to an ^{17}O in that site remote from the oxide surface. These data are consistent with a spin density of 0·5 on the $2p$ orbital of species I, 0·3 on species II and the remaining spin density 0·2 on the site originally occupied by the $(O_s^-)^+$ centre. $(O_3^-)_s$ centres in which both species I and II oxygens are the ^{17}O isotopes are observed at higher gain, indicating that there are three oxygens involved.

$(O_3^-)_s$ centres with no nearby proton may be produced by γ-irradiating MgO in the presence of O_2 or by the addition of O_2 to MgO which has been irradiated in vacuo. In these cases the molecular species is observed to interact with a V_s^- centre.

8.2. Optical studies of oxide surfaces

Nelson and Hale (1971) have characterized the surfaces of finely dispersed powders of MgO, CaO and SrO using the diffuse reflectance method. For powders exposed to air the optical reflectance spectra are characteristic of the absorption of bulk materials with large band gaps. However, powders evacuated and outgassed at high temperature show the presence of strong fluorescences which have been interpreted as due to direct band-gap processes involving oxygen $3s \rightarrow 2p$ transitions. Note that these 'effective' band gaps occur at lower photon energies than the known bulk band gaps. This effective band gap apparently results from the reduced interionic separation close to the powder surface.

The reflectance spectra of powders which have surface electronic states may be interpreted in terms of absorption by surface defects. For example MgO powder after γ-irradiation *in vacuo* is a blue-purple colour due to a broad absorption at $\lambda_{max} = 600$ Å. Parallel ESR and reflectance spectra show that this band is due to F_s^+ centres. The corresponding absorption spectrum in CaO consists of two broad bands at 6500 Å and 9500 nm. One might expect two transitions for F_s^+ centres since the strong axial distortion partially lifts the degeneracy of the excited '$2p$' like state. This is apparently confirmed by measurements on the F_s^+(H) centres, which also show broad absorption spectra with two peaks in each of the oxides MgO, CaO and SrO. The results are compared in table 22: it is surprising that these bands are shifted so far into the red in comparison with the bulk F^+ centres. In principle the '$1s-2p_{x,y}$-like' transitions are expected to be very close to the F^+ band, the '$1s-2p_z$ transition' being shifted to somewhat lower energy.

Table 22. Optical bands observed in powder reflectance studies (after Nelson and Hale 1971, Tench 1972). All transition energies are given in electron volts

Oxide		MgO	CaO	SrO
Band Gap	Bulk	8·7	7·7	6·7
	Powder	5·7	3·6	4·9
F^+ centre		4·95	3·6	3·0
F_s^+		2·05	1·90	—
			1·30	
$F_s^+(H)$ centre		{ 1·77	{ 1·65	{ 1·60
		{ 1·00	{ 0·91	{ 0·76
$(O_2^-)_s$		2·9	3·0	{ 3·0
				{ 1·7
$(O_s^-)^+$		2·0	—	—
$(O_3^-)_s$		2·95	—	—

The broad optical bands due to the adsorbed species $(O^-)_s^+$, $(O_2^-)_s^+$ and $(O_3^-)_s^+$ all lie close to one another, indicating that crystal field effects are relatively unimportant. The position of the $(O^-)_s^+$ absorption band is in good agreement with the reported absorption band of bulk V^- centres (Chapter 4).

Infra-red absorption spectra often depend quite markedly on the method of preparation and pretreatment of the oxide powder. There have been numerous studies of hydroxyl ions on MgO and CaO surfaces, and the principal spectroscopic features are well characterized. Broad bands due to physisorbed water molecules are observed *ca* 3500 cm^{-1}, whereas $Mg(OH)_2$ and $Ca(OH)_2$ absorb near 3700 cm^{-1}. On MgO powder isolated hydroxy-groups absorb at 3730 and 3610 cm^{-1}: the deuterated species give bands at 2760 and 2670 cm^{-1}. The particular conformation of the isolated hydroxide surface groups appears to be a matter of speculation. Detailed studies have been reported of surface hydroxy groups on CaO by Low *et al.* (1971). They find that a free OH$^-$ group adsorbs at 3695 cm^{-1} (2719 cm^{-1}), bound OH$^-$ absorbs at 3550 cm^{-1} (2625 cm^{-1}) and hydroxy groups associated with a brucite-like pseudolattice absorb at 3707 cm^{-1} (2734 cm^{-1}). (The values in parentheses are the frequencies observed for the corresponding deuteroxy groups.) The effects of evacuation treatments at temperatures in the range 300–1000°C have been studied by Faure (1970). Gregg and Ramsay (1970) have investigated adsorption of CO_2 on magnesium oxide powders: there appears to be a rapid physical adsorption and a slow chemisorption which initially results in the formation of first bidentate carbonate ions and then CO_3^{--} ions typical of $MgCO_3$. This latter reaction predominates if the temperature is raised above 100°C. The adsorption of CO on MgO surfaces also results in the formation of surface carbonate ion complexes (Kölbel et al. 1970).

Adsorption of pyridine on MgO and CaO surfaces has been observed (Tetryakov and Filiminov 1970) in the presence of isolated hydroxy-groups. The surface hydroxy-groups on both oxides are too basic to hydrogen bond to pyridine or benzonitrite. The surface species observed when methanol and ethanol are absorbed on MgO surfaces depend upon evacuation and heat treatment. At 25°C these alcohols are physisorbed on the oxide surface. Heating the reacting surface to 165°C produces infra-red bands characteristic of the formate and acetate species when methanol and ethanol respectively are adsorbed on the surface. At higher temperature surface methyl carbonate and ethyl carbonate are observed to be formed from methanol and ethanol respectively.

8.3. *Heterogeneous catalysis and chemical reactions on magnesium oxide surfaces*

There are numerous chemical interactions which are catalysed at oxide surfaces (e.g. hydrogen and oxygen exchange reactions, oxidation of hydrocarbons and inorganic molecules, polymerization reactions, decomposition reactions, isomerization, etc.). Such reactions have been studied using many oxide catalysts including ZnO, Al_2O_3, hydroxyapatite, V_2O_5, the zeolites and a host of mixed oxides. Much of this work is beyond the scope of the present treatise: however, the recent review by Scurrell (1974) will give the interested reader a comprehensive account of much of the current literature and an impression of the vast amount of work currently being undertaken in this area. Less work has been done on CaO, SrO and almost none on BaO. The work on MgO is important because it has in many cases given clear insights into the mechanisms of particular oxides and the species involved in these reactions. Frequently the use of ^{17}O has helped elucidate mechanisms especially where paramagnetic species such as V_s^- centres, $(O^-)_s^+$, $(O_2^-)_s^+$ and $(O_3^-)_s$ have been confirmed. For some considerable time the position of the $(O^-)_s^+$ centre on MgO and other oxide surfaces remained obscure although conductivity and chemisorption studies implied its presence in chemical reactions. The recent ESR results (Wong and Lunsford 1971, Tench et al. 1972) clearly demonstrate that some earlier assignments of ESR were incorrect.

Let us review first the mechanisms of formation of $(O^-)_s^+$ on MgO surfaces. As noted earlier this is accomplished by the decomposition of N_2O, presumably in consequence of the reactions

$$N_2O + e \to N_2O^- \to N_2 + O^-.$$

This reaction is well known in gas phase studies and is expected to occur readily on surfaces with good electron donor properties (Tench et al. 1972). This criterion is readily met with in MgO, since electrons trapped at discrete surface anion vacancies are readily produced by ultra-violet or γ-irradiation. The reacting species was actually the $F_s^+(H)$ centres (Wong and Lunsford 1971, Tench et al. 1972). ESR implicates the reaction

$$F_s^+(H) + N_2O \to N_2 + (O^-)_s^+.$$

Since the $F_s^+(H)$ is converted directly to $(O^-)_s^+$ it follows that the oxygen released in the dissociation reaction occupies the anion vacancy: this is signified by the excellent correlation of the initial concentration of $F_s^+(H)$ centres and final concentration of $(O^-)_s^+$ centres. The amount of released O^2 is negligible whereas the amount of released N_2 is about a factor of 3 larger than the $(O^-)_s^+$ concentration. Thus a significant concentration of other surface species involving oxygen must be produced, although their ESR spectra have not been measured. As we have already mentioned the $(O^-)_s^+$ centre may be present in at least three different environments on the powder surface differentiated by their g-values. The use of N_2O decomposition in evaluating catalytic activity is widely known. Transition metal ions dissolved in MgO are more active than when dissolved in ZnO. Further in MgO:CaO solid solutions it is found that the specific activity per cobalt ion is enhanced by increasing the dilution (Cimino and Pepe 1972). Cu^+ ions dispersed in the surface layers have also been shown to enhance the activity of MgO powders in decomposing nitrous oxide (Cordische et al. 1973).

The $(O^-)_s^+$ species appears to be especially reactive, much more so than the molecule ion $(O_2^-)_s^+$: reactions with O_2, H_2, CO_2 as well as several alcohols and propene (Tench et al. 1972). The reaction of $(O^-)_s^+$ with O_2 leading to the formation of $(O_3^-)_s$ is exothermic:

$$(O^-)_s^+ + O_2 \rightarrow (O_3^-)_s + Q$$

Q according to Tench (1972) is of order 1·5 eV. Thus (O_3^-) ions are relatively stable on oxide surfaces, although by heating to 400°C this ion does decompose to form $(O_2^-)_s^+$. In the presence of H_2, powders containing $(O^-)_s^+$ immediately turn blue due to $F_s^+(H)$ centre formation, although some spin concentration is lost because the measured $F_s^+(H)$ concentration is less than the original concentration. Surface reactions between $(O^-)_s^+$ and CO_2 lead largely to the formation of $(CO_3^-)_s$, although small amounts of $(CO_2^-)_s$ and $(O_2^-)_s$ may also be present.

Reactions of methanol, ethanol and isopropanol with $(O^-)_s^+$ are very similar, ESR spectra typical of the methoxy-, ethoxy- and isopropoxy-radicals adsorbed on the oxide surface being observed. A typical initial reaction is

$$CH_3OH + (O^-)_s^+ \rightarrow (CH_2OH)_s + (OH^-)_s.$$

No new ESR spectra are observed after the addition of propene. However, the characteristic $(O^-)_s^+$ spectrum is destroyed, indicating that the olefin has been oxidized (Tench et al. 1972).

Several authors have investigated hydrogen-deuterium equilibration over MgO powders. A study of the catalitic effect for the reaction

$$H_2 + D_2 \rightleftharpoons 2HD$$

indicates a correlation between the catalytic activity and the concentration of V_1 centres measured from the ESR spectrum (Lunsford 1964). The V_1 centres were introduced by 5 eV irradiation. The effect was pronounced in pellets that were degassed at about 565°K but no catalytic enhancement nor formation of V_1 centres were found for samples degassed at 775°K. The presence of appreciable concentrations of OH^- ions would appear to be necessary to provide charge compensation for the cation vacancies present. It would seem likely that a large fraction of the V centres were actually V_{OH} centres (§ 5). It was not clearly established whether only the V centres in the body of the crystallite are involved (Nelson et al. 1967); it is possible that the number of holes trapped at cation sites on the surface varies approximately in the same way as the concentration of V centres in the interior of the crystallites. Further studies will be required to clarify the mechanism.

A more recent study by Boudart *et al.* shows that the rate of equilibration at 77°K is affected only by surface protons. Pretreatment of the MgO powder *in vacuo* between 500–900°C could alter the surface reactivity by five orders of magnitude. A complex surface centre, thought to be OH^- adjacent to three O^- ions arranged in a triangular array on the (111) plane, has been associated with the enhanced catalytic activity. MgO has been observed to catalyse para-hydrogen and orthodeuterium conversions and $H_2 \rightarrow D_2$ equilibration both before and after γ-irradiation, the centre for catalytic activity being one of three paramagnetic surface centres (Zammith and Eley 1971).

Hydrogen-deuterium exchange in 3,3-dimethylbut-1-ene is also catalysed in the presence of MgO, but in the presence of D_2 replacement is confined to the three vinylic hydrogens (Kemball *et al.* 1972). If D_2O is used replacement of more than three hydrogens may occur due to slow isomerization of the reactant proceeding at high temperature via a carbonium ion intermediate. However, interconversion of 2.3-dimethylbut-1-ene and 2,3-dimethylbut-2-ene occurs rapidly at room temperature. The active sites on the surface of MgO were not identified, although presumably there is some similarity with but-1-ene isomerization (Baird and Lunsford 1972) over MgO powders. These authors concluded that O^{2-} ions located at cube corners were the catalytic centres, the concentration of which was maximized by heat treatment of the catalyst at 700°C. Hattori *et al.* (1972) find that for MgO and CaO catalysts of but-1-ene isomerization, the *trans*-isomer is formed at mainly acidic sites with the *cis*-isomer being formed at basic sites. The allyl carbonion appears to play an intermediary role. There is some similarity between these reactions and catalysis involving such oxides as Cr_2O_3 and SrO_2.

References

Abragam, A., and Pryce, M. H. L., 1950, *Proc. Phys. Soc.*, **63**, 409.
Abraham, M. M., Boatner, L. A., Lee, E. J., and Weeks, R. A., 1967, *Proc. of VIth Rare-Earth Research Conference*, Gatlinburg (unpublished).
Abraham, M. M., Boatner, L. A., Chen, Y., Kolopus, J. L., and Reynolds, R. W., 1971, *Phys. Rev.*, **B4**, 2853.
Abraham, M. M., Butler, C. T.. and Chen, Y., 1971, *J. Chem. Phys.*, **55**, 3752.
Abraham, M. M., Chen, Y., Boatner, L. A. and Reynolds, R. W., 1975 a, *Sol. State Comm.*, **16**, 1209.
Abraham, M. M., Chen, Y., Kolopus, J. L., and Tohver, H. T. (1972), *Phys. Rev.*, **5B**, 4945.
Abraham, M. M., Chen, Y., Lewis, J. T., and Modine, F. A. (1973), *Phys. Rev. B*, **7**, 2732.
Abraham, M. M., Chen, Y., and Unruh, W. P., 1973, *Phys. Rev. B*, **9**, 1842; 1975 b, *Phys. Rev. B*, **12**, 4766.
Abraham, M. M., Unruh, W. P., Chen, Y., 1974, *Phys. Rev. B*, **10**, 3539.
Adrian, F. J., 1957, *Phys. Rev.*, **107**, 488.
Allsop, A. L., Owen, J., Hughes, A. E., 1973, *J. Phys. C*, **6**, L337.
Ashby. M. F., and Brown, L. M., 1963, *Phil. Mag.*, **8**, 1083.
Austerman, S. B., 1964, *J. Nucl. Mater.*, **14**, 759.
Auzins, P., Orton, J. W., and Wertz, J. E., 1963, *Paramagnetic Resonance*, edited by W. Low (New York: Academic Press), **1**, 90.
Auzins, P., and Wertz, J. E., 1964, *Bull. Am. Phys. Soc.*, **9**, 707; 1967, *J. Phys. Chem., Ithaca*, **71**, 211.
Baird, M. J., and Lunsford, J. H., 1972, *J. Catalysis*, **26**, 440.

BARTRAM, R. H., SWENBERG, C. E., and FOURNIER, J. T., 1965, *Phys. Rev.* A, **139,** 491.
BARTRAM, R. H., SWENBERG, C. E., and LA, S. Y., 1967, *Phys. Rev.*, **162,** 759.
BATES, J. B., and WOOD, R. F., 1975, *Sol. State Comm.*, **17,** 201.
BENNETT, J. E., INGRAM, D. J. E., and SCHONLAND, D., 1956, *Proc. Phys. Soc.* A, **69,** 556.
BERSUKER, I. B., 1963, *Soviet Phys. JETP*, **16,** 933 and **17,** 836; 1965, *J. Theor. Exp. Chem.*, **1,** 1.
BESSENT, R. G., CAVENETT, B. C., and HUNTER, L. C., 1968, *J. Phys. Chem. Solids*, **29,** 1523.
BESSENT, R. G., and FELTHAM, P., 1967, *Phys. Stat. Sol.*, **25,** K 107.
BETHE, H., and VAN DER LARGE, F. C., 1947, *Phys. Rev.*, **71,** 612.
BIRD, B. D., OSBORNE, G. A., and STEPHENS, P. J., 1972, *Phys. Rev.* B, **5,** 1800.
BLAKE, W. J., GITELSON, H. A., and WERTZ, J. E., 1971, *J. Phys.* C (*Sol. State Phys.*), **4,** L261.
BOAS, J. F., HALL, T. P. P., and HUGHES, A. E., 1973, *J. Phys.* C., **6,** 1639.
BOATNER, L. A., REYNOLDS, R. W., ABRAHAM, M. M., and CHEN, Y., 1973, *Phys. Rev. Lett.*, **31,** 7.
BOUDART, M., DELBOUILLE, A., DEROUANE, E. G., INDOVINA, V., and WALTERS, A. B., 1972, *J. Amer. Chem. Soc.*, **94,** 6622.
BOWEN, D. H., 1963, *Trans. Br. Ceram. Soc.*, **62,** 771.
BOWEN, D. H., and CLARKE, F. J. P., 1963, *Phil. Mag.*, **8,** 1257; 1964, *Ibid.*, **9,** 413.
BOWEN, D. H., CLARKE, F. J. P., and WILKS, R. S., 1962, *J. Nucl. Mat.*, **6,** 148.
BOYD, C. A., RICH, D., and AVERY, E., 1947, *Rep. Congr. Atom. Energy Common.*, *U.S.*, MDDC 1508.
BRANSCOMB, L. M., 1962, *Atomic and Molecular Processes*, edited by D. R. Bates (New York: Academic Press).
BRIGGS, A., and BOWEN, D. H., 1968, *Symposium on Mass Transport in Oxides*, N.B.S. publication 296, 103.
BUBE, H., and STRIPP, K. F., 1952, *J. Chem. Phys.*, **20,** 193.
BUTLER, C. T., STURM, B. J., and QUINCY, R. B. (1971), *J. Cryst. Growth*, **8,** 197.
CABANNES-OTT, C., 1956, *C. r. hebd. Séanc. Acad. Sci.*, Paris, **242,** 355.
CARSON, J. W., HOLCOMB, D. F., and RUCHARDT, H., 1959, *J. Phys. Chem. Solids*, **12,** 66.
CASTNER, T. G., and KÄNZIG, W., 1957, *J. Phys. Chem. Solids*, **3,** 178.
CHASE, L. L., 1968, *Phys. Rev.*, **168,** 341.
CHENG, J. C., MANN, A., OSBORNE, G. A., and STEPHENS, P. J., 1972, *J. Chem. Phys.*, **57,** 4051.
CHEN, Y., and SIBLEY, W. A., 1969, *Phil. Mag.*, **20,** 217.
CHEN, Y., and SIBLEY, W. A., 1967, *Phys. Rev.*, **154,** 842.
CHEN, Y., KOLOPUS, J. L., and SIBLEY, W. A. (1969)a, *Phys. Rev.*, **186,** 865.
CHEN, Y., MAJOR, R. J., and MODINE, F., 1976, *Bull. Amer. Phys. Soc.*, **21,** 344.
CHEN, Y., SIBLEY, W. A., SRYGLEY, F. D., WEEKS, R. A., HENSLEY, E. B., and KROES, R. L., 1968 a, *J. Phys. Chem. Solids*, **29,** 863.
CHEN, Y., Trueblood, D. L., SCHOW, O. E., TOHVER, H. T., 1970, *J. Phys. Soc.*, **3,** 2501.
CHEN, Y., WILLIAMS, R. T., and SIBLEY, W. A., 1968 b, *Bull. Am. Phys. Soc.*, **13,** 415.
CHEN, Y., WILLIAMS, R. T., SIBLEY, W. A., 1969, *Phys. Rev.*, **182,** 960.
CHO, K., 1968, *J. Phys. Soc. Jap.*, **25,** 1372.
CIMINO, A., and PEPE, F., 1972, *J. Catalysis*, **25,** 362.
CLARKE, F. J. P., 1957, *Phil. Mag.*, **2,** 607.
CODLING, A. J., and HENDERSON, B., 1971 a, *J. Phys.* C, **1,** 1242; 1971 b, *Ibid*, **4,** 1409.
COFFMAN, R. E., 1968, *J. Chem. Phys.*, **48,** 609.
COFFMAN, R. E., LYLE, D. L., and MATTISON, D. R., 1968, *J. Phys. Chem., Ithaca*, **72,** 1392.
COLES, B. A., ORTON, J. W., and OWEN, J., (1960) *Phys. Rev. Lett.*, **4,** 116.
COMPTON, D. M. J., BRYANT, J. F., and CESENA, R. A., 1966, *Physics of Quantum Electronics* (McGraw-Hill).

COPELAND, W. D., and SWALIN, R. A., 1968, *J. Phys. Chem. Solids*, **29**, 313.
CORBETT, J. W., 1966, *Electron Irradiation in Semiconductors* (New York: Academic Press).
CORDISCHE, D., PEPE, F., and SCHIAVELLO, M., 1973, *J. Phys. Chem.*, **77**, 1240.
CRAWFORD, M. F., KELLY, F. M., SCHAWLOW, A. L., and GRAY, W. M., 1949, *Phys. Rev.*, **76**, 1527.
CULVAHOUSE, J. W., HOLROYD, L. V., and KOLOPUS, J. L., 1965, *Phys. Rev.* A, **140**, 1181.
DASGUPTA, S., 1966, *Br. J. Appl. Phys.*, **17**, 267.
DASH, W. C., 1953, *Phys. Rev.*, **92**, 68.
DAS, T. P., and DICK, B. G., 1962, *Phys. Rev.*, **127**, 1063.
DAVIDGE, R. W., 1967, *J. Mater. Sci.*, **2**, 339; 1968, *J. Nucl. Mater.*, **25**, 75.
DAVIES, J. J., 1968, *Phys. Lett.*, **28A**, 9.
DAVIES, J. J., SMITH, S. R. P., and WERTZ, J. E., 1969, *Phys. Rev.*, **178**, 608.
DAVIES, J. J., and WERTZ, J. E., 1969, *J. Mag. Res.*, **1**, 500; 1970, *J. Phys. Chem. Solids*, **31**, 2489; 1975, *J. Phys. C*, **8**, 1235.
DELBECQ, C. J., and PRINGSHELM. P., 1953, *J. Chem. Phys.*, **21**, 794.
DELBECQ, C. J., SMALLER, B., and YUSTER, D. H., 1958, *Phys. Rev.*, **111**, 1235.
DEUTSCHBEIN, O., 1932, *Ann. Phys.*, **14**, 712.
DEXTER, D. L., 1958, *Phys. Rev.*, **111**, 119.
DICK, B. G., 1966, *Phys. Rev.*, **145**, 609.
DOORN, C. Z. VAN, 1962, *Philips Res. Rep.* (Suppl. No. 4).
DRUMHELLER, J. E., 1964, *Helv. Phys. Acta*, **37**, 689.
DRUMHELLER, J. E., and RUBINS, R. S., 1964, *Phys. Rev.* A, **133**, 1099.
DURAN, J., MERLE D'AUBIGNÉ, Y., and ROMESTAIN, R., *J. Phys. C*, **5**, 2225.
DUVARNEY, R. C., and GARRISON, A. K., 1973, *Sol. St. Comm.*, **12**, 1235.
EDEL, P., HENNIES, C., MERLE D'AUBIGNÉ, Y., ROMESTAIN, R., and TWAROWSKI, Y., 1972, *Phys. Rev. Lett*, **28**, 1268.
EDEL, P., MERLE D'AUBIGNÈ, Y., and LOUAT, R., 1974, *J. Phys. Chem. Solids*, **35**, 67.
EISENSTEIN, A. S., 1954, *Phys. Rev.*, **93**, 1017.
ELEY, D. D., and ZAMMITH, M. A., 1971, *J. Catalysis*, **21**, 377.
ESCRIBE, C., and HUGHES, A. E., 1971, *J. Phys. C*, **4**, 2357.
EVANS, B. D., and KEMP, J. C., 1970, *Phys. Rev. B*, **2**, 4179.
EVANS, B. D., COMAS, J., and MALMBERG, P. R., 1972, *Phys. Rev. B*, **6**, 2453.
FAURE, M., 1970, *Bull. Foc. Chim.*, France, 69.
FEHER, 1964, *Phys. Rev.*, A, **136**, 145.
FEUCHTWANG, T. E., 1962, *Phys. Rev.*, **126**, 1616.
FITCHEN, D. B., 1968, *The Physics of Colour Centres*, edited by W. E. B. Fowler (New York: Academic Press).
FITCHEN, D. B., SILSBEE, R. H., FULTON, T. A., and WOLF, E. L., 1963, *Phys. Rev. Lett.*, **11**, 275.
FLETCHER, J. R., MARSHALL, F. G., RAMPTON, V. W., ROWELL, P. M., and STEVENS, K. W. H., 1966, *Proc. Phys. Soc.*, **88**, 127.
FOWLER, W. B., 1968, *The Physics of Colour Centres* (New York: Academic Press).
GAGER, W. B., KLEIN, M. J., and JONES, W. H., 1964, *Appl. Phys. Lett.*, **5**, 131.
GAMBINO, R. J., 1965, *J. Appl. Phys.*, **36**, 656.
GAMBLE, F. T., BARTRAM, R. H., YOUNG, C. G., GILLIAM, O. R., and LEVY, P. W., 1964, *Phys. Rev.* A, **134**, 589.
GANDY, H. W., 1958, *Phys. Rev.*, **111**, 764.
GARTON, G., HANN, B. F., WANKLYN, B. M., SMITH, S. H., 1972, *J. Cryst. Growth*, **12**, 66.
GERRITSEN, H. J., and SABISKY, E. S., 1962, *Phys. Rev.*, **125**, 1853.
GLASS, A. M., 1967, *J. Chem. Phys.*, **46**, 2080.
GLASS, A. M., and SEARLE, T. M., 1967, *J. Chem. Phys.*, **46**, 2092.
GLAZE, J. A., and KEMP, J. C., 1969, *Phys. Rev.*, **178**, 1502 and 1507.
GOURARY, B. S., and ADRIAN, F. J., 1957, *Phys. Rev.*, **105**, 1180.
GREGG, S. J., and RAMSAY, J. D., 1970, *J. Chem. Soc.* (A), , 2784.
GRIFFITHS, J. H. E., and ORTON, J. W., 1959, *Proc. Phys. Soc.*, **73**, 948.
GROVES, G. W., and FINE, M. E., 1964, *J. Appl. Phys.*, **35**, 3587.

GROVES, G. W., and KELLY, A., 1963, *Phil. Mag.*, **8,** 1437.
HALL, T. P., 1975, *J. Phys.* C, **8,** 1921; 1976, *J. Phys.* C, **9,** 1369.
HALLIBURTON, L. E., COWAN, D. L., BLAKE, W. B. J., and WERTZ, J. E., 1973 a, *Phys. Rev.* B, **8,** 1610.
HALLIBURTON, L. E., COWAN, D. L., and HOLROYD, L. V., 1973, *Sol. State Comm.*, **12,** 393, 1976, *Phys. Rev.*
HALLIBURTON, L. E., KAPPERS, L. A., COWAN, D. L., DRAVNIEKS, F., and WERTZ, J. E., 1973 b, *Phys. Rev. Lett.*, **30,** 607.
HAM, F. S., 1965, *Phys. Rev.*, **138A,** 1727; 1968, *Phys. Rev.*, **166,** 307.
HANSLER, R. L., and SEGELKEN, W. G., 1960, *J. Phys. Chem. Solids*, **13,** 124.
HARDING, B. C., 1967, *Phil. Mag.*, **16,** 1039.
HARTREE, D. R., HARTREE, W., and SWIRLES, B., 1939, *Phil Trans. Roy. Soc.*, **A238,** 229.
HARRIS, E. A., 1972, *J. Phys.* C, **5,** 338.
HARRIS, E. A., and OWEN, J., 1963, *Phys. Rev. Lett.*, **11,** 9.
HATTORI, H., YOSHII, N., and TANABE, K., 1972, *Proc. 5th Int. Congress on Catalysis* (North Holland, Amst.), **1,** 233.
HENDERSON, B., 1976, *J. Phys.* C, **9,** L579.
HENDERSON, B., 1964, *Phil. Mag.*, **9,** 153; 1966, *Br. J. Appl. Phys.*, **17,** 851; 1967, unpublished, 1968, *Symp. on Mass Transport in Oxides*, N.B.S. publication 296, 41.
HENDERSON, B., and BOWEN, D. H., 1971, *J. Phys.* C, **4,** 1487.
HENDERSON, B., BOWEN, D. H., BRIGGS, A., and KING, R. D., 1971 c, *J. Phys.* C, **4,** 1946.
HENDERSON, B., CHEN, Y., and SIBLEY, W. A., 1972, *Phys. Rev.*
HENDERSON, B., and GARRISON, A. K., 1973, *Adv. in Phys.*, **22,** 423.
HENDERSON, B., and HALL, T. P. P., 1967, *Proc. Phys. Soc.*, **90,** 511.
HENDERSON, B., and KING, R. D., 1966, *Phil. Mag.*, **13,** 1149; 1967, *J. Phys. Paris* (Supp. C4), **28,** 75; 1968, *Phys. Stat. Sol.*, **26,** K147.
HENDERSON, B., KING, R. D., and STONEHAM, A. M., 1968, *J. phys. Soc.* C (*Proc. phys. Soc.*), **1,** 586.
HENDERSON, B., KOLOPUS, J. L., and UNRUH, W. P., 1971 b, *J. Chem. Phys.*, **55,** 3519.
HENDERSON, B., and SIBLEY, W. A., 1971, *J. Chem. Phys.*, **55,** 1276.
HENDERSON, B., STOKOWSKI, S. E., and ENSIGN, T. C., 1969, *Phys. Rev.*, **183,** 826.
HENDERSON, B., and TOMLINSON, A. C., 1969, *J. Phys. Chem. Solids.* **30,** 1801.
HENDERSON, B., WERTZ, J. E., HALL, T. P. P., and DOWSING, R. D., 1971 a, *J. Phys.* C, **4,** 107.
HENRY, C. H., SCHNATTERLY, S. E., and SLICHTER, C. P., 1965, *Phys. Rev.* A, **137,** 583.
HENRY, C. H., and SLICHTER, C. P. in *The Physics of Colour Centres*, Ed. W. Beall Fowler (Academic Press, New York, 1968). Chapter 6.
HENRY, M. O., LARKIN, J. P., and IMBUSCH, G. F., 1976, *Phys. Rev.*, **B13,** 1893.
HENSLEY, E. B., and KROES, R. L., 1968, *Bull. Am. Phys. Soc.*, **13,** 420.
HIBBEN, J. H., 1937, *Phys. Rev.*, **51,** 530.
HICKMAN, B., and WALKER, D. G., 1965, *Phil. Mag.*, **11,** 1101.
HIRSCHFELDER, J. O., 1938, *J. Chem. Phys.*, **6,** 795.
HÖCHLI, U., MÜLLER, K. A., and WYSLING, P., 1965, *Physics Lett.*, **15,** 5.
HOLTON, W. C., and BLUM, H., 1962, *Phys. Rev.*, **125,** 89.
HORST, R. B., ANDERSON, J. H., and MILLIGAN, D. E., 1962, *J. Phys. Chem. Solids*, **23,** 157.
HOSKINS, R. H., and VAN STEENWINKEL, R., 1964, *Phys. Rev.* A, **133,** 490.
HUGHES, A. E., 1966, D.Phil. Thesis, University of Oxford (unpublished).
HUGHES, A. E., 1967, *J. Phys. Paris* (Suppl. C4), **28,** 55.
HUGHES, A. E., 1968, *J. Phys. Chem. Solids*, **29,** 1461.
HUGHES, A. E., 1970, *J. Phys.* C, **3,** 627.
HUGHES, A. E. (1973), *J. de Physique*, Supp. 11–12, **34,** 515.
HUGHES, A. E., and HENDERSON, B., 1972, *Defects in Crystalline Solids I*, edited by J. H. Crawford and L. M. Slifkin (Plenum New York).

HUGHES, A. E., and PELLS, G. P., 1972, *J. Phys.* C, **5**, 2543.
HUGHES, A. E., and PELLS, G. P., 1974, *J. Phys.* C, **7**, 3997.
HUGHES, A. E., and PELLS, G. P., 1975, *J. Phys.* C, **8**, 3703.
HUGHES, A. E., PELLS, G. P., and SONDER, E., 1972, *J. Phys.* C, **5**, 709.
HUGHES, A. E., POOLEY, D., RAHMAN, H. U., and RUNCIMAN, W. A., 1967, A.E.R.E. R5604.
HUGHES, A. E., and RUNCIMAN, W. A., 1965, *Proc. Phys. Soc.*, **86**, 615; 1967, *Ibid.*, **90**, 827; 1969, *J. Phys. C (Proc. phys. Soc.)*, **2**, 37.
HUGHES, A. E., and WEBB, A. P., 1973, *Sol. State Comm.*, **13**, 167.
IBERS, J. A., and SWALEN, J. D., 1962, *Phys. Rev.*, **127**, 1914.
IZEN, E. H., MAZO, R. M., and KEMP, J. C., 1973, *J. Phys. Chem. Solids*, **34**, 1431.
IMBUSCH, G. F., 1964, Ph.D. Thesis, Stanford University.
IMBUSCH, G. F., SCHAWLOW, A. L., MAY, A. D., and SUGANO, S., 1965, *Phys. Rev.* A, **40**, 830.
JOHNSON, B. P., and HENSLEY, E. B., 1967, *Bull. Am. phys. Soc.*, **12**, 411.
JOHNSON, B. P., and HENSLEY, E. B., 1969, *Phys. Rev.*, **180**, 931.
JOHNSON, P. D., 1954, *Phys. Rev.*, **94**, 845.
JONES, G. D., 1967, *Phys. Rev.*, **155**, 259.
KANE, E. O., 1951, *J. Appl. Phys.*, **22**, 1214.
KÄNZIG, W., 1955, *Phys. Rev.*, **99**, 1890.
KÄNZIG, W., and COHEN, M. H., 1959, *Phys. Rev. Lett.*, **3**, 509.
KAPLIANSKII, W., 1964, *Optics Spectrosc. Wash.*, **16**, 329 and 557.
KAPPERS, L. A., DRAVNIEKS, F., and WERTZ, J. E., 1972, *Sol. St. Comm.*, **10**, 1265.
KAPPERS, L. A., KROES, R. L., and HENSLEY, E. B., 1970, *Phys. Rev.*, **B1**, 4151.
KAZUMATA, Y., OZAWA, K., and NAKAGAWA, M., 1965, *Phys. Rev.*, **19**, 529.
KEMBALL, C., LEACH, H. F., SKUNDRIC, B., and TAYLOR, K. C., 1972, *J. Catalysis*, **27**, 416.
KEMP, J. C., 1963, *Bull. Am. phys. Soc.*, **8**, 484.
KEMP, J. C., and NEELEY, V. I., 1963 a, *Phys. Rev.*, **132**, 215; 1963 b, *J. Phys. Chem. Solids*, **24**, 332.
KEMP, J. C., CHENG, J. C., IZEN, E. H., and MODINE, F. A., 1969, *Phys. Rev.*, **179**, 818.
KEMP, J. C., ZINIKER, W. M., and GLAZE, J. A., 1966, *Physics Lett.*, **22**, 37; 1967 a, *Proc. Br. Ceram. Soc.*, **9**, 109.
KEMP, J. C., ZINIKER, W. M., GLAZE, J. A., and CHENG, J. C., 1968, *Phys. Rev.*, **171**, 1024.
KEMP, J. C., ZINIKER, W. M., and HENSLEY, E. B., 1967 b, *Physics Lett.*, A, **25**, 43.
KINCHIN, G. S., and PEASE, R. S., 1955, *Rep. Prog. Phys.*, **18**, 1.
KING, R. D., 1967, Ph.D. Thesis, University of Reading (unpublished).
KING, R. D., and HENDERSON, B., 1966, *Proc. phys. Soc.*, **89**, 153; 1967, *Proc. Br. ceram. Soc.*, **9**, 63.
KIRKLIN, P. W., AUZINS, P., and WERTZ, J. E., 1965, *J. Phys. Chem. Solids*, **26**, 1067.
KLICK, C., 1972, *Defects in Crystalline Solids I*, in J. H. Crawford and L. M. Slifkin (Eds.), (Plenum, New York).
KNEUBÜHL, F. K., 1960, *J. Phys. Chem., Ithaca*, **33**, 1074.
KÖLBEL, H., RALEK, M., and JIRU, P., 1970, *Z. Naturforsch.*, **25a**, 670.
KOLOPUS, J. L., and HOLROYD, L. V., 1965, *Phys. Stat. Sol.*, **8**, 711.
KORTÜM, G., 1969, *Reflexions-spektroscopie* (Springer-Verlag, Berlin).
KRUPKA, D. C., and SILSBEE, R. H., 1966, *Phys. Rev.*, **152**, 816.
KUNATH, D., and REKLAT, A., 1971, *Z. Chem.*, **11**, 361.
LANZL, F., VON DER OSTEN, W., and WAIDELICH, W., 1966, NATO Summer School on " Colour Centres in Solids ", Ghent.
LARKIN, J. P., IMBUSCH, G. F., and DRAVNIEKS, F., 1973, *Phys. Rev.*, **B7**, 495.
LAX, M., 1952, *J. Chem. Phys.*, **20**, 1752.
LIDIARD, A. B., 1967, *Proc. Br. Ceram. Soc.*, **9**, 1.
LINDNER, R., and PARFITT, G. D., 1957, *J. Chem. Phys., Ithaca*, **26**, 182.
LOW, M. J. D., TAKEZAWA, N., and GOODSELL, A. J., 1971, *J. Colloid Interface Sci.*, **37**, 422.

Low, W., 1957, *Phys. Rev.*, **105**, 801; 1958, *Phys. Rev.*, **109**, 247, 256; 1960, *Solid State Physics*, edited by Seitz and Turnbull (Academic Press).
Low, W., and Rubins, R. S., 1963, *Paramagnetic Resonance*, edited by W. Low (New York: Academic Press), **1**, 79.
Low, W., and Suss, J. S., 1963, *Physics Lett.*, **7**, 310.
Low, W., and Weger, M., 1960, *Phys. Rev.*, **118**, 1119, 1130.
Low, W., and Offenbacher, E. L., 1965, *Solid State Physics*, edited by Seitz and Turnbull (New York: Academic Press), **17**, 135.
Loudon, R., 1964, *Proc. Phys. Soc. (London)*, **84**, 379.
Ludlow, I. K., 1966, *Proc. phys. Soc.*, **88**, 763.
Ludlow, I. K., 1968, *J. Phys. C (Proc. phys. Soc.)*, **1**, 1194.
Ludlow, I. K., 1970, Ph.D. Thesis, University of London (unpublished).
Ludlow, I. K., and Runciman, W. A., 1965, *Proc. phys. Soc.*, **86**, 1081.
Lunsford, J. H., 1964, *J. Phys. Chem., Ithaca*, **68**, 2312; 1967, *J. Chem. Phys.*, **46**, 4347.
Lunsford, J. H., and Jayne, J. P., 1965, *J. Phys. Chem., Ithaca*, **69**, 2182; 1966, *J. Chem. Phys.*, **44**, 1487, 1492.
Lüty, F., 1968, *Physics of Colour Centres*, W. B. Fowler (Ed.), (Academic Press, New York).
Macfarlane, R. M., 1967, *J. Chem. Phys.*, **47**, 3066.
Mann, A. J., and Stephens, P. J., 1973, *Phys. Rev. B*, **9**, 863.
Mann, A. J., Stephens, P. J.—paper on CO^{2+}.
Mann, K. E., and Holroyd, L. V., 1968, *Phys. Stat. Sol.*, **28**, K27.
Mann, K. E., Holroyd, L. V., and Cowan, D. L., 1969, *Phys. Stat. Sol.*, **33**, 391.
Manson, N. B., 1971 a, *Phys. Rev.*, **4**, 2645; 1971 b, *Phys. Rev.*, **4**, 2656.
Marguglio, J., and Kim, S., 1975, *J. Chem. Phys.*, **62**, 1497.
Markham, J. J., 1959, *Rev. Mod. Phys.*, **31**, 956; 1966, *F-centres in the Alkali Halides* (New York: Academic Press).
Martin, D. G., 1968, *Proc. phys. Soc. (J. phys. C)* **1**, 333.
Matkin, D. I., and Bowen, D. H., 1965, *Phil. Mag.*, **12**, 1209.
McGeehin, P., and Henderson, B., 1974, *J. Phys. C*, **7**, 3988.
McMahon, D. H., 1964, *Phys. Rev. A*, **134**, 128.
Merle d'Aubigné, Y., *Proc. 18th Ampere Cong.*, 1974, **1**, 39.
Merle d'Aubigné, Y., 1976, *Defects and Their Structure in Non-Metals* (ed. Henderson and Hughes, Plenum).
Merle d'Aubigné, Y., and Roussel, A., 1971, *Phys. Rev.*, **3B**, 1421.
Mergerian, D., and Marshall, S. A., 1962, *Phys. Rev.*, **127**, 2015.
Miles, G. D., 1964, *Prog. Appl. Mater. Res.*, **5**, 26.
Miles, G. D., Clarke, F. J. P., Henderson, B., and King, R. D., 1965, *Proc. Br. ceram. Soc.*, **6**, 325.
Miles, G. D., and Lang, A. R., 1965, *J. appl. Phys.*, **36**, 1803.
Mitoff, S. P., 1962, *J. Chem. Phys.*, **36**, 1383.
Modine, F., 1972,
Modine, F. A., 1973, *Phys. Rev. B*, **7**, 1574.
Modine, F. A., Chen, Y., Major, R. W., and Wilson, T. M., 1976, *Phys. Rev. B*, **14**, 1739.
Morgan, C. S., and Bowen, D. H., 1967, *Phil. Mag.*, **16**, 165.
Mort, J., Lüty, F., and Brown, F. C., 1965, *Phys. Rev. A*, **137**, 566.
Morton, J. R., 1964, *Chem. Res.*, **64**, 453.
Neeley, V. I., 1964, Ph.D. Thesis, University of Oregon.
Nelson, C. M., and Pringshelm, P., 1948, Argonne Nat. Lab. Rep. 4232.
Nelson, R. L., and Hale, J. W., 1971, *Trans. Far. Soc.*, **52**, 77.
Nelson, R. L., and Tench, A. J., 1964, *J. Chem. Phys.*, **40**, 2736.
Nelson, R. L., Tench, A. J., and Harmsworth, B. J., 1967, *Trans. Faraday Soc.*, **63**, 1427.
Nelson, R. L., Tench, A. J., and Wilkinson, R. W., 1965, *Proc. Br. Ceram. Soc.*, **5**, 181.
Nishi, M., Fujita, T., Fujii, A., and Kato, S., 1971, *J. Phys. Soc. (Jap.)*, **31**, 612.

O'BRIEN, M. C. M., 1964, *Proc. Roy. Soc.*, **281**, 323; 1969, *Phys. Rev.*, **187**, 40; 1971, *J. Phys.* C, **4**, 2524; *J. Phys.* C, **9**, 3153.
O'BRIEN, M. C. M., 1976, *J. Phys.* C, **9**, 3153.
O'MARA, W. C., DAVIES, J. J., and WERTZ, J. E., 1969, *Phys. Rev.*, **179**, 816.
O'MARA, W. C., and WERTZ, J. E., 1970, *Sol. State Comm.*, **8**, 807.
O'MARA, W. C., 1969, Ph.D. Thesis, University of Minnesota. Unpublished.
ÖPIK, U., and PRYCE, M. H. L., 1957, *Proc. R. Soc.* A, **238**, 425.
ORTON, J. W., AUZINS, P., GRIFFITHS, J. H. E., and WERTZ, J. E., 1961, *Proc. Phys. Soc.*, **78**, 554.
ORTON, J. W., AUZINS, P., and WERTZ, J. E., 1960 a, *Phys. Rev.*, **119**, 1691; 1960 b, *Phys. Rev. Lett.*, **4**, 128.
OVENALL, D. E., and WHIFFEN, D. H., 1961, *Molec. Phys.*, **4**, 135.
OWEN, J., 1961, *J. Appl. Phys.* (Suppl.), **32**, 213.

PAKE, G. E., 1962, *Paramagnetic Resonance* (New York: Benjamin).
PAPPALARDO, R., WOOD, D. L., and LINARES, R. C., Jr., 1961, *J. Chem. Phys.*, **35**, 1460.
PECKHAM, G., 1967, *Proc. Phys. Soc.*, **90**, 657.
PERIA, W., 1958, *Phys. Rev.*, **112**, 423.
PIERCE, C. B., 1964, *Phys. Rev.* A, **135**, 83.
POOLE, C. P., 1967, *Electron Spin Resonance: A Comprehensive Treatise on Experimental Techniques* (New York, Interscience).
POOLEY, A., 1966, *Proc. Phys. Soc.*, **87**, 245, 257.
RALPH, J. E., and TOWNSEND, M. G., 1970, *J. Phys.* C, **3**, 8; 1968, *J. Chem. Phys.* **48**, 149.
RATINEN, H., 1972, *Phys. Stat. Sol.* (*a*), **12**, 175.
REDINGTON, R. W., 1952, *Phys. Rev.*, **87**, 1066.
REILING, G. W., and HENSLEY, E. B., 1958, *Phys. Rev.*, **112**, 1106.
REYNOLDS, R. W., BOATNER, L. A., CHEN, Y., and ABRAHAM, M. M., 1974,
REYNOLDS, R. W., CHEN, Y., BOATNER, L. A., and ABRAHAM, M. M., 1972, *Phys. Rev. Lett.*, **29**, 18.
RIUS, G., COX, R., PICARD, P., and SANTIER, C., *1970 Comp. Rend.*, **271**, 724.
RIUS, G., and HERVE, A., 1974, *Sol. St. Comm.*, **15**, 399 and 421.
ROMESTAIN, R., MERLE D'AUBIGNÉ, Y., 1971, *Phys. Rev.*, **B4**, 2009.
ROSE, B. H., and COWAN, D. L., 1974, *Sol. St. Comm.*, **15**, 775.
RUNCIMAN, W. A., 1965, *Proc. Phys. Soc.*, **86**, 625.

SAKSENA, B. D., and PANT, L. M., 1955, *J. Chem. Phys.*, **23**, 5.
SAUNDERSON, D., and PECKHAM, G. W., 1971, *J. Phys.* C, **4**, 2009.
SCHAWLOW, A. L., 1962, *J. Appl. Phys.*, (Suppl.), **33**, 395.
SCHAWLOW, A. L., WOOD, D. L., and CLOGSTON, A. M., *Phys. Rev. Lett.*, **3**, 544.
SCHIRMER, O. F., 1971, *J. Phys. Chem. Solids*, **32**, 499.
SCHIRMER, O. F., 1973, *J. Phys.* C, **6**, 300.
SCHIRMER, O. F., 1976, *Z. Physik*, **B24**, 235.
SCHIRMER, O. F., KOIDE, P., and REIK, H. G., 1974, *Phys. Stat. Sol.*, **B62**, 385.
SCHOENBERG, A., SUSS, J. T., LUZ, Z., and LOW, W., 1974, *Phys. Rev.*, **B9**, 2047.
SCHULMAN, J. H., and COMPTON, W. D., 1962, *Colour Centres in Solids* (New York: Pergamon Press).
SCURRELL, M. S., 1974, *Ann. Rev. Chem. Soc.*, **71**, 87.
SEITZ, F., 1946, *Rev. Mod. Phys.*, **18**, 384; 1954, *Ibid.*, **26**, 7.
SHARP, J. V., and RUMSBY, D., 1973, *Radiation Effects*, **17**, 65.
SHIREN, N. S., 1961, *Phys. Rev. Lett.*, **6**, 188.
SHUSKUS, A. J., 1963, *J. Chem. Phys.*, **39**, 849; 1964, *Ibid.*, **40**, 1602.
SIBLEY, W. A., and CHEN, Y., 1967, *Phys. Rev.*, **160**, 712.
SIBLEY, W. A., NELSON, C. M., and CHEN, Y., 1968, *Bull. Am. Phys. Soc.*, **13**, 415.
SILSBEE, R. H., 1965, *Phys. Rev.* A, **138**, 180.
SMITH, D. R., and TENCH, A. J., 1968, *Chem. Comm.* p. 1113.
SMITH, D. Y., 1965, *Phys. Rev.* A, **137**, 574.
SMITH, S. R. P., AUZINS, P. V., and WERTZ, J. E., 1968, *Phys. Rev.*, **166**, 222.
SMITH, S. R. P., DRAVNIEKS, F., and WERTZ, J. E., 1968, *Phys. Rev.*, **178**, 471.

SONDER, E., and SIBLEY, W. A., 1972, in 'Defects in Crystalline Solids', Ed. J. H. Crawford, Jr. and Slifkin, L. M., Plenum, N.Y.
SOROKIN, P. P., GELLES, I. L., and SMITH, W. V., 1958, *Phys. Rev.*, **112,** 1513.
SOSHEA, R. W., DEKKER, A. J., and STURTZ, J. P., 1958, *J. Phys. Chem. Solids*, **5,** 23.
SPAETH, J. M., 1966, *Z. Physik*, **192,** 107.
SPROULL, R. L., BEVER, R. A., and LIBOWITZ, G., 1953, *Phys. Rev.*, **92,** 77.
SPROULL, R. L., DASH, W. C., TYLER, W. W., and MOORE, A. W., 1951, *Rev. Sci. Instrum.*, **22,** 410.
SROUBEK, Z., 1961, *Czech. J. Phys. B*, **11,** 639.
STETTLER, J. D., SHATAS, R. A., and TANTON, G. A., 1965, *Bull. Am. phys. Soc.*, **10,** 108; 1966, *Physics Lett.*, **23,** 70.
STOKES, R. J., 1962, *Trans. metall. Soc. A.I.M.E.*, **224,** 1227.
STONEHAM, A. M., 1967, *A.E.R.E.*, Rep. No. R5510.
STONEHAM, A. M., and HAYES, W., 1975, *Crystals with the Fluorite Structure*, in W. Hayes (Ed.), (Oxford University Press).
STURGE, M. D., 1963, *Phys. Rev.*, **130,** 639.
SURPLICE, N. A., 1966, *Br. J. appl. Phys.*, **17,** 175.
SYMONS, M. C. R., 1962, *Adv. Chem. Sci.*, **36,** 76.
TAKEDA, T., and WATANABE, A., 1966, *J. phys. Soc. Japan*, **21,** 267.
TANIMOTO, D. H., 1966, Ph.D. Thesis, University of Oregon.
TANIMOTO, D. H., and KEMP, J. C., 1966, *J. Phys. Chem. Solids*, **27,** 887.
TANIMOTO, D. H., ZINIKER, W. M., and KEMP, J. C., 1965, *Phys. Rev. Lett.*, **14,** 645.
TENCH, A. J., 1972, *J.C.S. Farad. I*, **68,** 1181.
TENCH, A. J., and NELSON, R., 1966, *J. Chem. Phys.*, **44,** 1714; 1967 a, *Proc. Phys. Soc.*, **92,** 1055; 1967 b, *Trans. Faraday Soc.*, **63,** 2254; 1968, *J. Colloid. Sci.*, **26,** 364.
TENCH, A. J., and NELSON, R., 1968, *J. Colloid Sci.*, **26,** 364.
TENCH, A. J., LAWSON, T., and KIBBLEWHITE, J. F. J., 1972, *J.C.S. Faraday I*, **68,** 1169.
TO, K. C., STONEHAM, A. M., and HENDERSON, B., 1969, *Phys. Rev.*, **181,** 1237.
TOHVER, H. T., HENDERSON, B., CHEN, Y., and ABRAHAM, M. M., 1972, *Phys. Rev.*, **5B,** 3276.
TOMLINSON, A. C., and HENDERSON, B., 1969, *J. Phys. Chem. Solids*, **30,** 1793.
TORRENS, I. MC., CHADDERTON, L. T., and MORGAN, D. V., 1966, *J. Appl. Phys.*, **37,** 2395.
TRET'YAKOV, N. E., and FILIMINOV, V. N., 1970, *Kinetics and Catalysis*, **11,** 815.
UNRUH, W. P., and CULVAHOUSE, J. W., 1967, *Phys. Rev.*, **154,** 861.
UNRUH, W. P., CHEN, Y., and ABRAHAM, M. M., 1973 a, *Phys. Rev. Lett.*, **30,** 446 and 1973 b, *J. Chem. Phys.*, **59,** 3284.
VENABLES, J. D., 1963, *J. Appl. Phys.*, **34,** 293.
WALLIS, R. F., HERMAN, R., and MILNES, H. W., 1960, *J. molec. Spectrosc.*, **4,** 51.
WALTER, G. K., and ESTLE, T., 1961, *J. Appl. Phys.*, **32,** 1854.
WARD, W. C., 1965, Ph.D. Thesis, University of Missouri, Columbia.
WARD, W. C., and HENSLEY, E. B., 1965, *Bull. Am. Phys. Soc.*, **10,** 307.
WEBER, H., 1951, *Z. Phys.*, **30,** 392.
WEIGHTMAN, P., and HALL, T. P. P., 1973, *J. Phys. C*, **6,** 1292.
WEIL, J. A., and HECHT, H. G., 1963, *J. Chem. Phys.*, **38,** 281.
WELCH, L. S., HUGHES, A. E., PELLS, G. P., and SCHOENBERG, A., 1976, 2nd Europhysical Topical Conference on 'Lattice Defects in Ionic Crystals', Berlin.
WERTZ, J. E., and AUZINS, P., 1957, *Phys. Rev.*, **106,** 484; 1965, *Ibid*, **139,** A, 1645; 1967, *J. Phys. Chem. Solids*, **28,** 1557.
WERTZ, J. E., AUZINS, P., GRIFFITHS, J. H. E., and ORTON, J. W., 1959, *Discuss Faraday Soc.*, **28,** 136.
WERTZ, J. E., AUZINS, P., WEEKS, R. A., and SILSBEE, R. H., 1957, *Phys. Rev.*, **107,** 1535.
WERTZ, J. E., and BOLTON, J., 1972, *Electron Spin Resonance* (McGraw Hill, New York).
WERTZ, J. E., and COFFMAN, R. E., 1965, *J. Appl. Phys.*, **36,** 2959.

Wertz, J. E., Hall, L. C., Hegelson, J., Chao, C. C., and Dykoski, W. S., 1967, *Interaction of Radiation with Solids*, edited by A. Bishay (New York: Plenum Press).
Wertz, J. E., Orton, J. W., and Auzins, P., 1961, *Discuss. Faraday Soc.*, **30,** 40; 1962, *J. Appl. Phys.* (Suppl.), **33,** 322.
Wertz, J. E., Saville, G., Auzins, P., and Orton, J. W., 1963, *J. Phys. Soc. Japan*, Suppl. II, **18,** 305.
Wertz, J. E., Saville, G. S., Hall, L., and Auzins, P., 1964, *Proc. Br. Ceram. Soc.*, **1,** 59.
Wilks, R. S., 1967, *A.E.R.E. Rep.* No. R5543.
Williams, F. E., 1962, *Phys. Rev.*, **126,** 70.
Williamson, W. B., Lansford, J. H., and Naccache, C., 1971, *Chem. Phys. Lett.*, **9,** 33.
Wilmshurst, T., 1967, *Electron Spin Resonance Spectrometers* (London: Hilger).
Wolga, G. J., and Tseng, R., 1964, *Phys. Rev. A*, **133,** 1563.
Wong, N., and Lunsford, J. H., 1971, *J. Chem. Phys.*, **55,** 3007.
Wood, R. F., 1970, *Phys. Stat. Sol.*, **42,** 849.
Wood, R. F., and Wilson, T. M., 1975, *Sol. St. Comm.*, **16,** 545.
Woodgate, G. K., 1970, *Elementary Atomic Structure* (McGraw Hill, London).
Woods, J., and Wright, D. A., 1955, *Proc. Phys. Soc. B*, **68,** 566.
Zamitt, H. A., and Eley, D. P., 1971, *J. Catal.*, **21,** 377.

Author Index

Abragam A., 17, 35, 44
Abraham M. M., 58, 60, 65, 67, 74
Adrian F. J., 19, 78, 99
Allsop A. L., 79, 82
Anderson P. W., 133
Ashby M. F., 130
Auzins P., 38, 48, 52, 68, 135, 137

Baird M. J., 144
Bartram R. H., 62, 70, 100, 108
Bates J. B., 102
Bennett H. S., 100
Bersuker J. B., 44
Berezin W. B., 106
Bessent R., 8, 10, 12, 76, 87, 89, 111
Bethe H., 99
Bird B. D., 55
Blake W. J., 65
Blum H., 19
Boas J. F., 103
Boatner L. A., 44
Boudart M., 144
Bowen D. H., 3, 4, 64, 75, 128, 131
Bowler R. S., 70
Boyd C. A., 85
Briggs A., 4, 65, 131
Brown L. M., 130
Bube H., 51, 54
Butler C. T., 44

Cabannes-Ott C., 65
Carson J. W., 8, 10, 87, 135
Castner T. G., 55
Chase L. L., 36
Chen Y., 9, 11, 44, 51, 53, 58, 60, 67, 85, 93, 96, 119, 126
Cho K., 92
Cimino A., 143
Clarke F. J. P., 3, 4, 5, 75, 85, 127, 131
Codling A. J., 38
Coffman R. E., 44, 54
Coles B. A., 41
Compton D. M. J., 54
Compton W. D., 2
Corbett J. W., 2, 18
Cordische D., 143
Cowan D., 58, 70
Cox R., 68
Crawford M. F., 99
Culvahouse J. W., 79, 99

Dash W. C., 8
Das T. P., 81
Davidge R. W., 4
Davies J. J., 33, 35, 38
Delbecq C. J., 21, 69
Deutschbein O., 51
Dexter D. L., 87
Dick G. B., 81
Dravnieks F., 50, 64
Drumheller J. E., 45

Duran J., 92
DuVarney R. C., 58

Edel P., 96, 102
Eley D. D., 144
Escribe C., 92
Estle T., 135
Evans B. D., 93, 95, 126

Faure F. M., 149
Feher G., 45, 49
Feltham P., 8
Feuchtwang T. E., 63, 73
Filiminov V. N., 142
Fine M. E., 6
Fitchen D., 21, 24
Fletcher J. R., 51, 54
Fowler W. B., 2

Gager 8
Gambino R. J., 3
Gamble F. T., 63
Gandy H. W., 53
Garrison A. H., 58, 71, 74, 77, 126
Garton G., 3
Gerritzen H. J., 43
Glass A. M., 52, 53, 58, 64
Glaze J. A., 123
Gourary B. S., 19, 78, 99
Gregg S. J., 141
Griffiths J. H. E., 53
Groves G. W., 6, 130

Hale J. W., 133, 140
Hall, T. P. P., 38, 51, 58, 68, 92, 103, 105, 111, 127
Halliburton L. E., 58, 74, 79, 81
Ham F. S., 90
Hansler R. L., 53
Hartree D. R., 62
Harris E. A., 41
Hattori H., 144
Hayes W., 102
Henderson, B., 9, 12, 35, 38, 41, 44, 59, 62, 85, 90, 98, 102, 106, 118, 124
Henry G. H., 30
Henry M. O., 53
Hensley E. B., 8, 9, 11, 99
Hervé A., 68, 70
Hecht G., 136
Hickman B., 131
Hirschfelder J. O., 123
Höchli U., 43, 53, 55
Holroyd L. V., 45
Holton W. C., 19
Hoskins R. H., 5
Hughes A. E., 11, 21, 24, 27, 55, 58, 62, 69, 79, 90, 92, 102, 115

Ibers J. A., 136
Izen E. H., 58, 70
Imbusch G. F., 51

Jayne J. P., 137
Jette N., 71
Johnson B. P., 9, 86, 96
Jones G. D., 41

Kane E. O., 87
Känzig W., 55
Kaplianskii A., 25, 27, 115
Kappers L. A., 69, 86, 93, 95
Kazumata A., 115
Kelly A., 130
Kemball C., 144
Kemp J. C., 8, 9, 10, 45, 79, 82, 85, 88, 99
Kim Y. S., 39
Kinchin G. S., 13, 85, 127
King R. D., 9, 12, 83, 85, 111, 115, 117, 119, 122, 124
Kirklin P. W., 64
Klick C. C., 102
Kneubuhl F. K., 136
Kölbel H., 142
Kolopus J. L., 45, 61, 66
Kortüm G., 140
Kroes R., 9
Krupka D. C., 123
Kunath D., 135

Lanzl F., 21, 24
Larkin J., 53
Lax M., 22, 24
Low M. J. D.,
Low W., 7, 34, 41, 43, 48, 54
Ludlow I. K., 115, 117, 121, 122, 125
Lunsford J. H., 137, 139, 142, 143
Lüty F., 102

Macfarlane R. K., 53
Marguglio, 39
Mann A. K., 54
Mann K. E., 1, 78
Manson N. B., 55
Markham J. J., 22, 24
Martin D. G., 132
Marshall S. A., 139
Matkin D. I., 4
McGeehin P. M., 44, 55
Merle d'Aubigné Y., 30, 92, 98
Mergerian D., 139
Miles D. J., 2, 4
Modine F. A., 95
Morgan C. S., 132
Mort J., 87, 90
Morton J. R., 139

Neeley V. I., 9, 79, 82, 85, 99, 125
Nelson C. M., 85
Nelson R. S., 77, 108, 111, 133, 137, 140, 143
Nishi M.,
Norgett M. J., 106

O'Brien M. C. M., 44, 91, 95, 100
O'Mara W. C., 58, 65
O'Donnell K. P., 38, 52

Offenbacher E. L., 17
Öpik U., 43, 100
Orton J. W., 44, 48, 53
Ovenall D. E.,
Owen J., 41

Pake G. E., 32
Pant L. D., 53
Pappalardo R., 44, 53, 54
Peases, 12, 85, 127
Peckham G., 95, 114
Pells P., 55, 103
Pepe F., 143
Peria W., 68, 85
Pierce C. B., 115
Poole C. P., 30
Pooley D., 11
Pringsheim P., 21
Pryce M. H. L., 35, 43

Ralph J. E., 53, 55
Ramsay J. D., 141
Ratinen H., 55
Redington R. W., 10
Reiling G. W., 99
Reklat A., 135
Reynolds R. W., 44
Rius G., 68, 70
Romestain R., 93
Rose B. H., 70, 74, 96
Roussel A., 92
Rubins R. S., 41, 43, 45, 48
Rumsby D., 127
Runciman W. A., 25, 27, 90, 121

Saksena B. D., 53
Saunderson 95
Schawlow A. L., 51
Schirmer O. F., 65, 66, 68, 71, 74
Schnatterley S. E., 30
Schoenberg A., 43, 44
Schoemaker D., 71
Scholtz A., 133
Schulman J. H., 2
Scurrell M. S., 142
Searle T. M., 53, 58, 64
Seitz F., 2, 56
Sharp J. W., 127
Shiren N. S., 45
Shuskus A. J., 45, 48, 55
Sibley W. A., 6, 8, 11, 54, 60, 69, 102, 115
Silsbee R. H., 121
Slichter C. P., 30
Smith D. Y., 87, 90
Smith S. R. P., 45, 50
Sonder E., 6, 8, 102
Sorokin P. P., 45
Soshea R. W., 9, 68, 83, 85
Spaeth M. J., 81
Sproul R. L., 10
Sroubek Z.,
Stetler J. D., 115
Stokes R. J., 4, 6

Stoneham A. M., 102
Sturge M. D., 51, 53
Suss J., 44
Swalen J. D., 136
Symons M. C. R., 139

Takeda T.,
Tanimoto D. H., 45, 86, 103, 105, 111
Tench A. J., 77, 111, 133, 135, 139, 142
To K. C., 109, 111
Tohver H. T., 62, 69, 71, 75
Tomlinson A. C., 9, 35, 60, 67
Torrens I. Mc., 11
Tret 'Yakov N. E., 142

Unruh W. P., 58, 60, 66, 99

Van Dorn C. Z., 123
Venables J. D., 3, 4

Walker D. G., 128
Wallis R. F., 110

Walters G. K., 135
Ward W. C., 8, 9
Webb A. P., 93, 95
Weber H., 6, 9, 50
Weger M., 41, 45
Weightman P. W., 41, 104
Weil J. A., 136
Welch L., 104
Wertz J. E., 8, 9, 21, 33, 35, 48, 54, 55, 58, 64, 67, 77, 85, 98, 103, 108, 118, 137
Wilks R. S., 3
Wilmshurst T., 30
Wilson T. M., 85, 100, 102
Wong N., 139, 142
Wood R. F., 79, 83, 85, 100, 102
Woodgate G. K., 13
Woods J., 53, 54, 79
Wright D. A., 54

Zamitt D. D., 144
Ziniker W. M., 10

Subject Index

Absorption coefficient 20, 27, 85
Additive colouration 8, 10
  of BaO 10
  of CaO 9
  of MgO 8, 9
  of SrO 10
Active sites 140, 144

Adsorbed species 133–144
  absorption spectra of 141
  adatoms 134
  catalytic agents 134
  ESR of 135, 137–140
  $C_6H_5NO_2$ 137
  CO_2^- 137
  infrared absorption of 141
  O_2^- 137
  O_3^- 140
  optical studies of 140–142
  NO 138
  NO_2^{2-} 139

Aggregates
  of vacancies 7, 102, 105, 109
  dislocation loops 113
  electronic degeneracy 25, 27, 120
  electron microscopy 113, 5
  ESR of 104–111
  exchange effects 106
  nomenclature of 7, 68, 111
  orientational degeneracy 25, 27, 116
  polarized absorption 25, 120
  polarized emission 53
  uniaxial stress 25, 70, 104, 116, 120
  vacancy/impurity 51, 103, 64
  vibrational properties 114
  zero phonon lines 103, 116–126

Alkali halides 1, 2
  aggregate centres 7, 115
  ESR spectra
  F_3 centre in 121, 123
  V-band in 59
  X_2^- centres in 55, 71
  zero-phonon lines 21, 115

Alkaline earth oxides
  annealing studies 6, 50, 108, 112, 125
  cavity formation 131
  charge compensation 34, 36, 51
  crystal growth 3, 44
  defect structure 7, 55, 126
  dislocations in 3–5, 130
  heat treatment of 50, 108
  impurities in 3, 7, 126
  impurity luminescence 51–54
  impurity precipitate 4, 5
  F^+ centres in 75–95
  F centres in 95–100
  F_A centres in 102–104
  point interstitials 128
  surface defects 133–144
  voids in 132
  V-type centres 55–75

Barium oxide
  crystal growth 3
  F^+ centres in 78, 87, 100
Born–Oppenheimer approximation 22
Brillouin zone boundaries 113
  density of states at 114

Calcium Oxide
  F^+ centre in 77, 86, 89, 93, 100
  F_A^+ centre in 103
  F centre in 95
  impurities in 6, 35
  transition metal ions
  triplet state defects 96, 105, 111
  V-type centres in 55–75
Catalysis 133
Catalytic exchange 55, 143
Cavity information 131
Colour centres 18
Condon approximation 22
Configurational co-ordinate model 22,
  of F centre 101
Crystal field 17, 32
Crystal growth
  arc fusion 3, 44
  flux growth 3
  flux evaporation 3
  plasma torch 3
  ^{17}O enrichment 78
Crystals
  additively coloured 8
  imperfect 2, 3
  electron microscopy 5
  macroscopic defects 3, 131
  surface energies 133

Defects
  classification 7, 102, 133
  clustering 103
  Frenkel 127, 128
  on surfaces 133–145
  point interstitials 128
  Schottky 127
Diffusion 6
Dislocation 5, 113
  grown-in 6
  impurity decoration 3
  loops 113, 130, 131
  networks 3
Displacement energy 11, 12

Effective mass model 110
ELDOR of V^- centres 71
Electric field effects
  on V^- centres 70
  on zero phonon lines
Electron microscopy
  of cavities 131–2
  of dislocation lines 3, 5
  of dislocation loops 131, 132

of impurities 3, 5, 65
of precipitates 3, 6
of voids 132
Electron phonon interaction 22
Electronic states
 of defects 21, 123
 degeneracy of 25
ENDOR 16–20
 of trapped hole centres 57–59, 64–68, 73
 of F^+ centres 76, 79–82
 selection rule 16
ESR 13–20
 anisotropy 19
 double quantum transitions 45
 F^+ centre spectrum 76–79
 F^+(Mg) centre spectrum 104
 F^R centre spectrum 96–98
 forbidden transitions 45
 hyperfine structure 15, 18
 of $3d^1$ ions 32
 of $3d^3$ ions 35
 of $3d^5$ ions 38
 of $3d^6$–$3d^9$ ions 41–44
 of P^- centre 109
 of polycrystalline oxides 133–140
 pair spectra 39, 41, 67, 104
 rare earth ions 44
 selection rules 15
 spectrometer 30
 of V^- centres, 58–73

Excited states 22–99
 of F centre 95
 of F^+ centre 87, 99
 Jahn–Teller effect in 90, 92, 98
 spin-orbit coupling in 90
 triplet states 96
 vibrational interaction in 92, 100
 V^- centres 61, 70, 74
F_A(Mg)
 in CaO 104
F aggregate centres 102–126
 in CaO 102
 in MgO 112–123
F centres 2, 95
 Faraday rotation 97
 optical absorption 86
 luminescence 93, 96
 theory of 100, 102
 triplet states in 96
F^+ centre 2, 9
 ENDOR spectra 79–82
 excited states 87, 89
 luminescence 93
 magneto-optic studies 87
 optical absorption 82–92
 spin Hamiltonian 18, 76, 81

Faraday rotation 9
 of F^+ centres 87
 of F centres 97
 of aggregates, 123
$F \rightarrow F^+$ photoconversion 10, 86, 95

F_s^+ centres 136, 137, 141
F_s^+ (H) centres 139, 141, 142
 reaction with No 139
Fermi–Segre contact interaction 18, 79, 81, 104
 in F_A^+ centres 104
 in trapped hole centres 71, 73
Frenkel defects 127
 disorder 129
 pairs 128

Ham reduction factor 92
Huang–Rhys factor 24, 86, 90, 92, 100, 103
Hyperfine splitting 35
Hyperfine structure interaction 15, 16, 33, 39
 anisotropic 19, 71
 of F^+ centres 76–82
 of F_A^+ centres 104
 of P^- centres 109
 of trapped hole centres 58–73
Hyperfine tensor 18

Impurity 7
 adatoms 131–143
 hydroxide 3, 63
 in powders 135
 macroscopic effects 2, 6
 $[M]^0$ centres 68–75
 rare earth ions 44, 55
 transition metal ions 3, 30–55
Infrared absorption 63, 133, 141
Inhomogeneous broadening 113
Interstitial defects 68, 126, 131
Interstitial Loops 130
Ionic defects 7

Jahn–Teller effect 30, 34, 75, 123
 dynamical 90
 static 98
Jahn–Teller energy 75, 98

Lifetime 96, 98
Lithium, impurity 63, 65
$[Li]^0$ centre 68, 70, 73
 distortion effects 73

Macroscopic defects 3–7, 130–133
 growth parameters 128
Magnesium oxide
 adsorbed species 137
 $3d^n$ ions in 30–55
 electron-excess centres 75–102
 Pair spectra 105
 trapped hole centres 55–75
 surface defects 133–144
 vacancy aggregates 115–126
Microstructure 5
Moments of bandshape 24, 86, 96

Neutron irradiation 12, 58, 75, 112, 126–133
Nuclear collision 12
Nuclear magnetic resonance 16, 20

Optical absorption 20

of F^+ centres 82–87
of F_A^+ centres 103
of F centres 95
of surface centres 140
of trapped hole centres 72
Optical spectroscopy 13, 20
Orthogonalized wave functions 19, 81
Orthorhombic defects 26, 116–119
Oscillator strength 26, 23
  of F^+ centres 85
  of F centres 95, 96
  of trapped hole centres 68, 70, 75

Paramagnetic resonance
  of adsorbed species 137
  of optically excited states 36, 96
Phonon assisted transitions 114
Phonon modes at surfaces 133
Point ion lattice model 99
Polarized luminescence 13, 53, 104

Rare earth ions
  ESR spectra of 44
  luminescence of 55

Schottky defects 127, 128
Schrödinger equation 22, 110
Selection rules 13
  in e.s.r. 13, 15
  in ENDOR 16
  in optical spectroscopy 13
Spectroscopy
  high resolution 28
  low resolution 29
Spin Hamiltonian 16–20
Spin Hamiltonian parameters
  of $3d^n$ ions 34–42
  of rare earth ions 46–47
  of F^+ centres
  of trapped hole centres 72

Spin lattice relaxation 31
Spin-orbit interaction 16, 32–35, 61, 88, 98, 121
Spin-orbit interaction 32–35, 61, 88, 98, 121
Stokes shift 23, 86
Stress-induced dichroism 93, 96, 104, 120
Strontium Oxide 10, 78, 89
Surfaces
  band gap 133, 140
  defects on 133–144
  energies 133
  phonon modes near 133

Trapped hole centres 55–75
  ESR and ENDOR spectra 56–61
  nature of 61, 71
  optical properties of 68
Transition metal ions 31
  ESR spectra of 31–50
  luminescence properties 51–55
Trigonal defects 119–125
Tyndall scattering 9

Uniaxial stress 25, 90, 98, 104, 115–123

V centres 67
V^- centres 7, 55–59, 69, 74
V_{AL} centres 7, 58, 69
V_F centres 67, 69
V_{OH} centres 63, 69
Vacancy pairs 108–111

Wave functions 22, 23, 79

X-ray growth 130
X-ray topography 5

Zeeman effect 16, 21, 36
Zero-phonon lines 21–25